WAKEFIELD PRESS

Barossa Shiraz

Dr Thomas Girgensohn is a former Managing Partner of the Boston Consulting Group and an Australian company director with experience in a range of industries. He was educated in Germany with an MBA from Saarbrücken and a PhD in business from the University of Munich. Thomas has been collecting Australian wine for nearly 30 years, closely observing and following developments in the wine industry over this period. He currently publishes a wine blog, in which he shares his tasting experiences: 'Alontin's Australian wine reviews – and beyond' (www.australianwinereviews.blogspot.com).

BAROSSA Shiraz

Discovering the Tastes of the Barossa's Regions

Thomas Girgensohn

Wakefield Press

Wakefield Press
1 The Parade West
Kent Town
South Australia 5067
www.wakefieldpress.com.au

First published 2013

Cover design by Stacey Zass
Maps produced by John Frith, Flat Earth Mapping, based on information from the SA Department of Environment and Natural Resources
Text designed and typeset by Michael Deves, Wakefield Press
Printed in China through Red Planet Print Management

National Library of Australia Cataloguing-in-Publication entry

Author:	Girgensohn, Thomas, author.
Title:	Barossa Shiraz: discovering the tastes of the Barossa's Regions / Thomas Girgensohn.
ISBN:	978 1 74305 237 2 (paperback).
Notes:	Includes bibliographical references.
Subjects:	Syrah (Wine) – South Australia – Barossa Valley. Wine blending – South Australia – Barossa Valley. Wine and wine making – South Australia – Barossa Valley – History. Wineries – South Australia – Barossa Valley – History.
Dewey Number:	663.200994232

Contents

Foreword

When I read the manuscript of Thomas Girgensohn's book *Barossa Shiraz*, it reminded me immediately of Max Lake's *Classic Wines of Australia* (1966), a deeply personal and reflective narrative of the emerging Australian fine wine scene. This important reference book remains extremely relevant and is never far from my side.

Like *Classic Wines of Australia, Barossa Shiraz* is a personal exploration based on a belief that Australia is making great and important wine. For Thomas Girgensohn the quintessential expression of great Australian wine is Barossa Shiraz. Certainly this genre is the most successful performer on the Australian secondary wine market. Steeped in a wonderful historic Anglo-German narrative and anchored in one of the world's most beautiful landscapes, Barossa Shiraz is gorgeously seductive, lasting and compelling. It's one of the most delicious of all wines. Vineyard names like Hill of Grace, Kalimna and Stonewell evoke a strong and inimitable viticultural identity. Grant Burge, Charles Melton, Elderton, Henschke, Kaesler, Penfolds, Peter Lehmann, Rockford, Torbreck and Yalumba, amongst others, powerfully articulate the essence and personality of Barossa Shiraz. Yet what makes these wines so unique and expressive?

Thomas Girgensohn is a longstanding wine collector with a deep love and fascination of Barossa Shiraz. With his considerable experience in the consulting, business management and philanthropic world, he brings a fresh and insightful look at the present and future of Barossa Shiraz. His observations and intuitive tastings, based on extensive personal experience and systematic review are topical, sagacious and prescient. This book follows in the tradition of *Classic Wines of Australia* where the identity and aesthetic appeal of Barossa Shiraz is explored for the sake of understanding its potential. Dr Thomas Girgensohn's *Barossa Shiraz* is a remarkable and enlightened achievement.

Andrew Caillard, MW

Introduction

Mother Nature has a big influence on the flavour and structure of wine. Heatwaves shrivel berries, sandstorms split them, cold weather prohibits them from ripening.

There is now a substantially increased interest in understanding where wines – our most complex agricultural product – come from, and how the particular place where the grapes are grown (the *terroir*) influences and shapes the final product. This parallels the increasing interest of consumers in food in general.

A number of specific factors are playing a part in this increase in interest:

Internationally, Australian wine, Shiraz in particular, is mostly seen as a branded product, with no particular personality or character; most of it does not deserve a high price (there are some exceptions), and it does not create consumer loyalty. Without such loyalty, in difficult economic times customers stop buying these wines. This has happened over the last few years. Winemakers now realise that they need to differentiate their product, make it special and rare. Linking it clearly to the place where the wine comes from is an obvious way of doing this.

Many wineries are now embracing organic and biodynamic principles. They are still a minority, but their numbers are growing rapidly. The specifics of a place are reflected in vines through what they take up through their leaves and roots, so as the amount of chemicals sprayed reduces, the influence of climate will be more directly reflected by the vines. Equally, the more sustainable the practices used on the soil, the more clearly the vines will reflect their soil environment. Less interventionist winemaking techniques mean less covering up of the specific aspects of weather and soil that affect grapegrowing and winemaking. The resulting wines are more individual in terms of both flavour and structure.

A number of high-quality, medium-sized wineries are gaining in influence and reputation. These are mostly wineries that bottle wine from the vineyards on the estate. They are focusing on where their wine originates.

Many wineries have started to introduce single-vineyard wines. This broadens their range, allows higher pricing and links their wine to its *terroir*.

In 1993, Australia signed a trade arrangement with the European Community under which it would properly codify its wine regions. Over 70 regions across the continent are now registered. Still, many regions are so large that their tasting profile is not consistent. The quest to understand *terroir* is about finer distinctions. As the desire to understand the link between wine and *terroir* increases, it becomes clear that it has not been studied systematically here in the way it has been in Europe, in particular in France.

This book is about Barossa Shiraz. The Barossa is Australia's best known wine region and Shiraz its defining grape. It therefore makes sense to review the impact of *terroir* on Barossa Shiraz first. Shiraz from the Barossa has recently received its share of criticism. The most common complaints are that these wines are too big and alcoholic and cannot be enjoyed with food, that they do not have enough natural acidity, and that they do not mature well. This line of thinking can lead to the conclusion that this wine should not remain as prominent as it has been, and even that under conditions of global warming, Shiraz may not be the most suitable grape for the Barossa. However, the cold and wet 2011 vintage, plus changes in winemaking techniques, lead me to think this is a very premature conclusion.

Unfortunately, there seem, more and more, to be waves of fashion in the wine industry; they invariably ignore past achievements. We had the 'Anything But Chardonnay' wave, we now have an anti-Sauvignon Blanc wave, and Australian wine critics are not taken seriously if they do not condemn ripe and alcoholic Shiraz from the Barossa. There is no doubt that there are over-extracted high-alcohol Shirazes which are simply not enjoyable for most people. But there are also well-made Shirazes with ripe flavours which are appreciated by many consumers and which occupy a unique place in the wine world.

I have been collecting and drinking Barossa Shiraz for close to 30 years. I have been amazed at the differences between these wines, and I have increasingly felt that this was not only due to the winemaker's influence.

This book attempts, for the first time, to systematically examine the relationship between *terroir* and wine in Australia, focusing on Barossa Shiraz. Apart from reviewing material which covers aspects of this topic, I have interviewed many winemakers and viticulturists in the Barossa, and undertaken many tastings, in particular from the barrel, in order to establish differences between areas within the Barossa. I have identified 11 sub-regions of the Barossa Valley (plus Eden Valley), each of which generates different flavour and structure profiles for Shiraz. This creates an opportunity for winemakers to link their wines to *terroir*, and for consumers to gain a more differentiated appreciation of Barossa Shiraz.

In Chapter 1 I develop the main thesis of this book, argue for the focus on Barossa Shiraz in the first instance and define *terroir*. The second chapter explains the current state of the wine industry in the Barossa in terms of its history and its economic context. Each of the following chapters looks at a group of sub-regions. Each chapter describes the *terroir* of each of its sub-regions, lists the major Shiraz wineries of each, with brief descriptions, offers a typical tasting profile for each sub-region and highlights vineyards which I consider benchmarks for the sub-regions. Among these chapters are profiles of leaders in the industry, the people who make it all happen. The book is not an annual guide to Barossa Shiraz, but I have listed those Shirazes that I have especially enjoyed. The book concludes with a summary of the findings relating to the flavour and structure of Barossa Shiraz in the various sub-regions, and lists the tasks which still need to be carried out to complete the picture and make it more robust.

Writing this book has been a challenge, as it covers new ground (no pun intended). It would not have been possible without the help of many people. My thanks go first of all to my wife Ingeborg, who encouraged me to embark on this project, to John and Jan Goodall, who read the early drafts and made valuable structural suggestions, to Brian Walsh and Louisa Rose, who supported my project from the start and made relevant documents from the Barossa Grape and Wine Association available to me, to the wineries who invited me to do comparative tastings from barrel, to David Powell and Robert O'Callaghan for their specific insights into the Barossa, to Tony Kalleske, Alex Head and Jim Lumbers for reviewing my writing and to the numerous winemakers and grape-growers who gave me their time. A special thanks to Sarah Shrubb, my editor, who put up with a native German speaker and persisted with her corrections, and to John Frith from Flat Earth Mapping, who produced the sub-regional maps, based on Topographic Mapping Data supplied by the Customer Service Centre, Client Services, Department of Environment and Natural Resources, Government of South Australia. And finally a special thankyou to Michael Deves from Wakefield Press. Michael provided guidance and support through the production process by contributing valuable comments and advice.

This book expresses the personal views of the author. The text has been influenced by the sources read and listed as references. Quotes are specifically referenced and images attributed where appropriate. While all reasonable attempts at factual accuracy have been made, the author accepts no responsibility for any errors contained in the book. However, any correction would be appreciated and should be forwarded to the author or publisher.

Thomas Girgensohn
Sydney, August 2012

The beautiful Barossa
[Photo: Dragan Radocaj]

1 Shiraz and *terroir* in the Barossa

Suddenly *terroir* is everywhere. Winemakers talk about it, wine critics talk about it, and even wine consumers talk about it. There is also an emphasis, in all this talk, on organic and biodynamic principles. Some of this is fashionable spin: 'keeping it simple', 'letting nature do its thing'. There is no doubt, however, that vineyards themselves are now being acknowledged as a crucially important part of the making of wine, rather than merely the source of the base ingredient. At the same time, though, there has been no systematic review of the impact of *terroir* in Australia. What is it? What does it do and does Australia have it? There are no simple answers to these apparently simple questions.

I have studied the impact of *terroir* on Shiraz in the Barossa for the last two years. This has involved talking to many winemakers and viticulturists, studying maps and carrying out structured tastings, in particular from barrels made by the same winemaker but from fruit of different vineyards. The result? I believe the impact of *terroir* is significant. This is great news for winemakers and consumers alike. Winemakers who wish to emphasise *terroir* can produce wine that is clearly distinguishable by the vineyard/s it is sourced from, and therefore rarer, and consumers can systematically explore different aspects of Barossa Shiraz.

The Barossa is Australia's most famous wine region. In 1996, the winemakers and grapegrowers defined the Barossa Zone, and a year later, within it, the Barossa Valley Region and the Eden Valley Region. The naming can be confusing, as the term 'Barossa Valley' is sometimes used to mean the whole Zone, and at other times more specifically the Region.

In the authoritative classification of fine wine by Langton's, Australia's leading wine auction house, 23 per cent of the listed wines come either entirely or in majority from the Barossa Zone, and according to Wine-Ark, a leading Australian wine club and wine storage business, Barossa wines are also the most collected. (The detailed outcome is somewhat dependent on the allocation of Penfolds wines.)

Internationally, wines from the Barossa are the most highly rated. The 1990 Penfolds Grange – the current vintages of Grange are predominantly sourced from the Barossa – was selected by the prestigious American *Wine Spectator* magazine as the Number 1 wine in the world for 1995.

The Barossa is also the home of many unique vineyards. The oldest known Cabernet Sauvignon vineyard in the world is in the Barossa, and some of the oldest Shiraz and Grenache vineyards are also found there.

The Barossa Zone, and within it the Barossa Valley (green) and the Eden Valley (yellow)

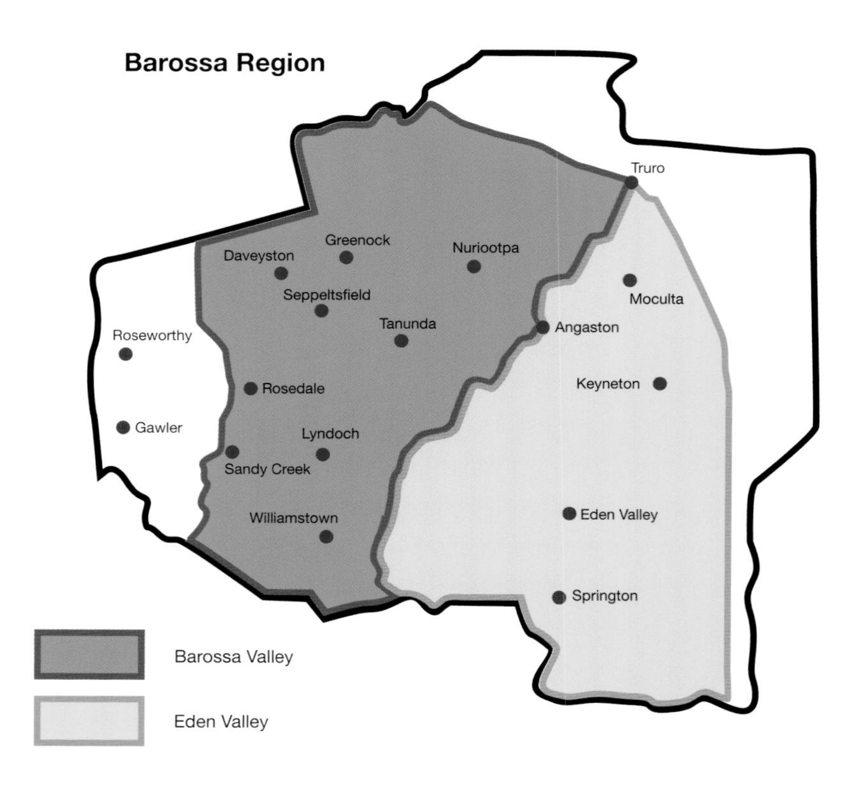

Given the importance of the region in the wine world, it is surprising that so little detail is known and documented about the Barossa. Many wine guides list its wineries, but there is little discussion about regional and sub-regional differences. It can be argued that it took the French several hundred years to come up with their classifications, but California, for example, which has a young wine industry, has already taken significant steps to define its wines' place of origin. It is this lack of detail that has stimulated the writing of this book.

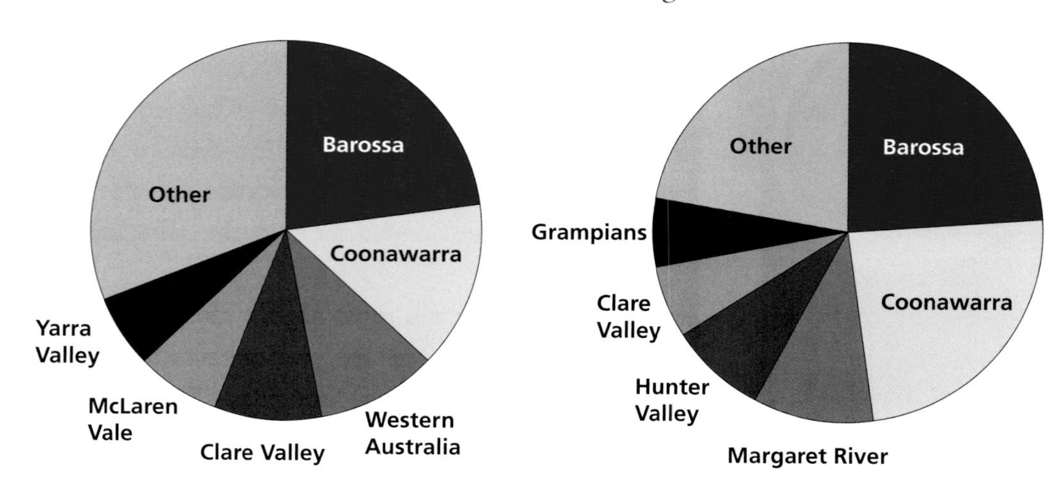

Percentage of wines in Langton's Wine Classification 2010

Percentage of wines collected according to Wine-Ark: Australia's Most Collected Wines 2009

The importance of sub-regions

Winemakers of the Barossa have stressed, in my discussions with them, the importance of 'brand Barossa'. And while this positioning was no doubt crucial in the development of the industry in the 1980s (see below), it will not be sufficient for the future. 'Australian wine' has increasingly, over the last 15 years, been seen as synonymous with 'cheap wine', and no distinctions have been made between Australian wines. Not only has the amount of wine exported fallen as a result, but the dollar value per litre has also fallen. These developments do not just threaten the industry at large; in particular they threaten high-quality and low-yielding areas such as the Barossa, which cannot compete on a cost basis. One response to this, in particular by smaller wineries, has been to increase flavour extraction, oak and alcohol, backed by high praise and ratings from American wine critic Robert Parker. Unfortunately, many of these wines have offered little individual character, so there was little lost for the consumers who dropped these often expensive wines during the 2008/09 Global Financial Crisis.

Increasingly, branded wine, which pays little attention to its origin, can be scaled up – in terms of production amounts – almost at will. This approach is running into headwind from wine writers and informed consumers alike. The best response has been a complete turnaround: a focus on bringing out the uniqueness of every wine instead of on making all wines taste the same.

The big opportunity in this quest for uniqueness is the *terroir*, or 'sense of place', which is present in the wine. 'Brand Barossa' will no longer be enough as a claim to uniqueness. The situation here parallels that in France, where generic Bordeaux or Burgundy can only be sold at very low prices.

The case for

The search is on for a higher level of specificity or uniqueness. At a minimum, this has to be established at the level of sub-regions. The potential value is clear: if a wine from one sub-region is in fact different from wine from any other sub-region, then it is a rarer wine. If its character is desirable, then its collectability increases.

There is quite a lot of talk about sub-regions in the Barossa at present. The Barossa Grape & Wine Association, representing the grapegrowers and wineries of the region, is carrying out a detailed long-term project to identify regional and other differences between Barossa wines. The number of single-vineyard wines, and wines with 'place' names is increasing. Some wineries, such as Chateau Tanunda or Dan Standish, either have established or are in the process of establishing a series of wines from different sub-regions. A number of wineries have registered trademarks referring to specific locations.

The definition of sub-regions will give more meaning to single-vineyard

Mengler Hill

Towards Rowland Flat

wines. It will also increase the understanding of blended wines and the approach which has been taken to create them. The benefit is simply that vines from a particular sub-region are made into wine which is specifically linked to its region's characteristics in scent, flavour and structure – and none of these can be duplicated elsewhere.

The case against

Two main arguments are put forward against the identification of sub-regions, and against using them in winemaking and in marketing.

The first is that significantly more than half of Barossa premium wine is blended. For these wines and their winemakers, sub-regions would in fact be detrimental if they were to lead to a prohibition of blending.

However, the discussion of sub-regions is not a discussion of the virtues or otherwise of blending wine, and there are of course arguments for and against blending. First, blending is a way to create and maintain a particular style of wine across different vintages and their variations. Leading producers in the Barossa also blend for other reasons. They understand the different contributions fruit from various regions can make to a wine, and they use blending to increase a wine's complexity and completeness (see The Blending of RunRig).

But what is blending, anyway? If a company separates two blocks in a vineyard and bottles its best grapes from one separately and then combines the remaining grapes from that block with those from the other block, is this a blended wine? One well-known winemaker harvests his 'home block' at four different times in order to reduce over-ripening risk and to balance ripeness and freshness. He has the opportunity to vary these four components and does so from year to year. The wine is designated 'single vineyard'.

The second issue is about appellations and regulations. Many winemakers point out that the freedom to blend different grapes, to blend between regions, and generally be quite free in the approach to winemaking, has led to a lot of innovation in Australia, and has given winemakers an advantage over their counterparts in Europe.

It has to be said that the French AOC (vins d'Appellation d'Origine Controlée) is often misunderstood, and its benefits in relation to maintaining quality standards (in vine density and maximum yields, for example) not sufficiently recognised. However, when I argue the case for sub-regionality, I am not arguing for increased regulation.

In summary, the introduction of sub-regions should not restrict blending or increase regulation. However, if it leads to increased differentiation, it will be of interest to the consumer and of value to growers and winemakers.

Near Seppeltsfield

Shiraz as the focus

Winemakers have many choices in the Barossa, just as they do in other parts of Australia. They can produce single-vineyard wines or they can blend. They can use any grape introduced to Australia. As a result, there is incredible variety in the Barossa, much more than in European wine regions. It is necessary to create an 'apples for apples' comparison in order to examine the validity of sub-regions for the Barossa Valley.

I have decided to base this investigation on Shiraz wines. There are a number of reasons for this: First, Shiraz is planted more widely than any other grape in the Barossa.

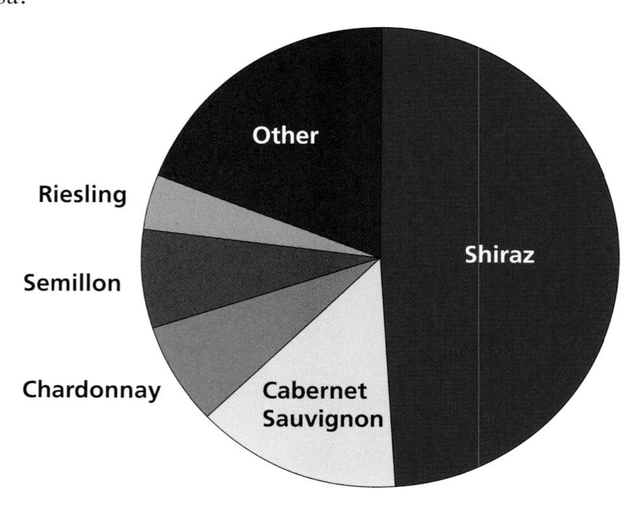

Grapevine Planting in the Barossa Valley

Also, Shiraz is grown widely in every area of the Barossa Valley.

Second, most of the acclaimed Barossa wines are Shiraz. Third, the Barossa Grape & Wine Association is basing its project about regional differences in the Barossa on Shiraz, at least initially. And finally, Shiraz adapts to its environment perhaps more readily and with more flexibility than many other grape varieties. It can be ripe and jammy, it can be tannic and chewy or peppery. Its flavours go from wild berries to plum to earthy to meaty characteristics. If there are differences in the environment, they should show in the flavour and structure of Shiraz.

I am not attempting in this book to define sub-regions for Eden Valley. While it can be argued that the area is even more complex than the Barossa Valley in terms of its climate, topography and soil, there simply are not enough plantings, certainly not of Shiraz, to allow a definitive sub-division. However, I will describe different areas of Eden Valley in Chapter 8.

History of Shiraz vines in the Barossa

The initial Shiraz cuttings that arrived in the Barossa came from New South Wales. It is important to look at the history of the grape, as the Barossa has a number of the oldest and most significant Shiraz vineyards in the world.

John Macarthur was the first who imported a significant number of different grape varieties to Australia, in 1817. He established a vineyard and vine nursery on the Nepean River near Penrith in 1820. His son William describes these early days in his 'Letters on the Culture of Vine'. Only a few grapes in this experimental vineyard survived, and Shiraz grapes were not among them. However, Shiraz cuttings from here were taken to the Hunter Valley in 1830, where George Wyndham, the founder of Wyndham Estate, planted the first commercial Shiraz vineyard in Australia.

In 1833, James Busby landed in Sydney with a collection of vines. They included Shiraz, or Scyras, as he called it. It was part of the private collection he had assembled on his travels through the main wine-producing areas of France and Spain. These vines are likely to have been cuttings from the village of Hermitage in France, regarded as the centre of quality for the Shiraz grape. The vines were initially planted in the Sydney Botanic Gardens. They did not survive there and were uprooted in 1859.

Luckily, however, way back in 1830 the Macarthurs had established a new and larger vineyard at Camden, and Shiraz cuttings from the Botanic Gardens were planted there in the 1830s. Some cuttings, including from Shiraz, were also sent to the Adelaide Botanic Gardens in 1839 by A.H. Davis. From here 'thousands of cuttings were spread throughout South Australia'.[1] Other growers also brought Shiraz vines to the Barossa later. Records are sparse, but Joseph Gilbert, the Englishman who established Pewsey Vale, reports vines being imported from the United Kingdom in 1847.[2] Incidentally, the latter are, according to Stephen and Prue Henschke, the source of the Mount Edelstone vines. But given that we have had Shiraz vines in the Barossa from early in the 1840s, it is reasonable to assume that the oldest Barossa Shiraz vines stem from the original transfer of the Busby Hermitage vines.

The Shiraz vines imported from the mid 19th century on have never been exposed to phylloxera, the pest which destroyed most vineyards in France – and many in Australia, but not South Australia – at the end of the 19th century. The European Shiraz vines have since been grafted onto resistant rootstocks, whereas the oldest Barossa vines are still the 'original' ones.

1 G.R. Gregory, 'Development and Status of Australian Viticulture', in B.G. Coombe and P.R. Dry, *Viticulture*, Vol. 1, Adelaide, Australian Industrial Publishers, 1988.

2 Ebenezer Ward, *The vineyards and orchards of South Australia*, Adelaide, printed and published at the offices of the *Advertiser* and *Chronicle*, 1862.

These old vines cannot be traced back to one 'mother' cell: they are polyclonal. The first clonal selections – vines relating back to one particular cell – in relation to Barossa Shiraz were made by Harry Tulloch at the Barossa Research Station in the late 1950s; they were not distributed to industry until the 1970s. The focus of Tulloch's work was on developing clones that were free from virus and would have an increased yield. Clones of other varieties were imported, but Shiraz was not, because the variety was very available. The material created at the research station then was rated highly, and a number of clones were taken back to France. In the initial panel tastings during this time, there were differences in the descriptions of the wines, but the panel could not identify differences in quality.

In the last 10 years, clonal development, in particular at Yalumba and Penfolds, has focused on quality rather than yield increase. There are now clones which show big differences. The most perplexing finding, according to Peter Dry, Associate Professor at the School of Agriculture and Wine, University of Adelaide, was that Eden Valley clones, when grown in the Barossa, showed characteristics similar to those of same clones when grown in Eden Valley. The current explanation is that clones may develop certain characteristics in a particular mesoclimate and, after some years, not lose them, or not lose them for a period of time.

These clones have not yet influenced the older benchmark vineyards, and most important wines described in this book are based on vines grown before clonal selection became commercially available. The conclusion is therefore that identified flavour and structure differences cannot be related to clonal differences, but must have other reasons. However, the future may show significant clonal influences as well.

Next I will discuss the place of sub-regionality and *terroir* in general and deal with some common preconceptions and misconceptions.

Nature v nurture

The concept of sub-regions is only relevant if *terroir* has a systematic influence on the scent, flavour and structure of wine. The term '*terroir*' is quite ill-defined and often misunderstood. Many use a narrow definition, such as temperature, and point out that wines from warmer areas taste bigger and richer than those from cooler areas. Others relate it to 'terrain' or 'territory', and often focus on soil composition. It is then assumed that the soil can directly influence the character of wine by transferring flavour components – such as chalk or slate, for instance – to the grape. However, there is no scientific evidence that this phenomenon, also called *goût de terroir*, exists. Wine is unlikely to taste of the soil it is grown in. All that vine roots need to take up from the soil is water and dissolved mineral ions, which are basically flavourless. An up-to-date summary on this topic is provided by David Farmer, a geologist who has been studying sub-regionality in the Barossa Valley for some time.

Soil is important, though, for these reasons:

1. It determines drainage, which is a critical factor in the ripening process of grapes. In Europe, the best *terroirs* appear to be those with free drainage and water tables that are high enough to supply water to roots on a regular, but not abundant basis.
2. It has an impact on how the sunlight is reflected back on the vines. Darker soils create more heat radiation. The large slab-like pebbles in the vineyards of Châteauneuf du Pape or the Gimblett gravels in Hawkes Bay, New Zealand are often quoted as examples of soils that soak up the heat during the day and radiate it back to the vines at night. John Gladstones (2011) believes that a narrower temperature range through the day has a positive influence on wine quality.
3. Darker and warmer soils accelerate grape development and ripening, which is still important in warm climate regions.
4. Mineral ions may alter plant metabolism and thus influence the growth of vines. Indirectly these ions may transfer certain flavours of the soil to wine by changing the vine's metabolism, but this is not proven. European Riesling producers, in particular, believe that they can taste the flavours of soil in the Riesling. After extensive tasting, I could make a similar case for Barossa Shiraz grown on soil with a heavy ironstone component.

Other significant aspects of *terroir* are the climate (temperature, rainfall and sunshine hours, predominantly) and the topography, including altitude and aspect. Wind and cloud cover can also be important.

In this book, *terroir* is defined in the following way:

> *Terroir* is the sum total of environmental factors influencing the growth of the grapevine and its grapes.

Vines are rooted in a particular spot. The grapes are produced from what is in the earth and from the light in the sky above it: this is what is meant by *terroir*.[3]

When *terroir* is used as a tasting term, it can mean the unique characteristics of a wine made from a particular piece of land.

The other major factors influencing wine are viticulture and winemaking. Viticulture is anything to do with vineyard management – density of planting,

3 For a detailed and comprehensive analysis of the impact of *terroir* on wine see John Gladstones: *Wine, Terroir and Climate Change*, Adelaide, Wakefield Press, 2011.

clone selection, pruning, picking time, etc. Winemaking is anything relating to decisions made in the winery – maceration period, the use of chemicals, the use of barrels and oak, etc. There are a lot of assertions made regarding which of these three factors is the most important:

1. We are just custodians of Mother Nature;
2. The wine is made in the vineyard;
3. The winemaker's decisions fashion the wine.

The fact that these are debated indicates in itself that there is no clear answer. *Terroir*, viticulture and winemaking all contribute significantly. High yields and interventionist winemaking will reduce the impact of *terroir*. The *terroir* influence on wines is growing as the use of organic and biodynamic principles increases. This is because chemical treatment reduces the impact of *terroir*. The less grapes are cultivated by chemicals, the lower the yield, and the more gentle and natural the winemaking process, the more the winemaking disappears and the land takes centre stage. The movement towards letting nature 'do its thing' will allow more uniqueness in the scent and flavour of grapes. The fact that the Barossa has phylloxera-free vineyards may add a further element of uniqueness and interest.

In addition, relative contributions will vary from vintage to vintage. In Australia, winemaking and branding of wine have received the most attention, but they may not be the most important factors in the long term, and particularly not for premium wines.

The French believe that a specific *terroir* will affect a wine in a specific way, irrespective of viticultural practices and winemaking techniques. Cistercian monks from the Côte d'Or discovered early in the first millennium that certain vineyards delivered the best grapes year after year. They concluded that it was due to the soil, and they started to grade the quality of the land into a 'cru' classification that is still in existence today. The power of this notion is reflected in land prices. If one were able to buy 1 hectare of good-quality land in Burgundy, one would have to part with more than A$3 million. In Australia, where special sites are less recognised, the value might be A$85,000–100,000.

The *terroir* concept is not universally accepted. Some regard it as a clever marketing tool, a way of saying French wine cannot be replicated and is therefore unique. Others go further and argue that it is an excuse for bad winemaking. Wine which does not taste much of fruit, but more of dull pasteboard, is labelled as 'showing particular *terroir*'.

Some sceptics point out that on many occasions foreign soils have been imported to Bordeaux, or that Château Petrus, one of its most heralded wineries, sits on pretty flat land that lacks gravel, which is regarded as desirable and high-quality soil in Bordeaux. On the other hand, Château Petrus gets quite a lot of moisture, but the swelling clay crushes the rootlets and prohibits its vines from absorbing too much of that water. In reality, the natural conditions which form the viticultural potential are complex and very intertwined.

However, skilled wine tasters can pick the region wines originate from pretty much without fail. There are characteristics in wine that are specific to regions.

Other sceptics don't believe the *terroir* concept can yet be applied to Australia. The argument is that Australia does not have enough history in growing wine to have extracted the specifics of a place. All major grape varieties are pretty much grown everywhere. There is no specialisation. I would counter that one only has to taste wine from Leeuwin's Chardonnay vineyard, Cullen's Cabernet vineyard or Henschke's Hill of Grace to know that unique *terroir* exists in Australia. In the words of Professor of Soil Science Robert White: 'the impact of *terroir* is incontrovertible through taste. You only have to drink good wine from specific vineyards to see that. Proving it scientifically, though, is very hard.'

Then there is the argument against *terroir* when it is discussed in relation to Shiraz: Shiraz, it is said, is a grape which can grow anywhere and the specifics are due to winemaking techniques. On the other hand, winemakers have pointed out to me that Shiraz, as a variety, is actually more sensitive to climate than other red varieties, including Cabernet Sauvignon and Tempranillo. I will review this further below, but a taste of, for example, the Blonde and Brune brands by Guigal, which are made from grapes from the opposite sides of the river Soave, will demonstrate that *terroir* can have a major influence on the final wine. In fact, Rhône winemakers have said that soil type is the most important factor in making wine in the Rhône Valley, ahead of vine age, clone, etc. If one tastes different Shiraz wines from the same winemaker, produced in the same way, from different vineyards, as I have done many times, the wines can taste vastly different.

The *terroir* concept is just as relevant for Australia as it is for France or any other country, and it is just as relevant for Shiraz as for any other grape. Different *terroirs* can be, at a minimum, different starting points for wines; on another view, they can be the key to a wine's uniqueness.

What, then, are the major aspects of the *terroir* influence? The following list is an attempt to capture the major factors, although no doubt this could be debated at length. When I think about *terroir*, I think about:

- climate (temperature and temperature range, rainfall, humidity);
- topography (altitude, aspect, slope); and
- soil.

Some of these factors are interrelated.

What makes this concept difficult to grasp is that not all factors are equally relevant in all regions. For example, some regions may be defined predominantly by soil type, some by altitude, some by the age of plantings.

One further complication is that nature tends not to be homogenous and neat. If you look at a micro level, a particular vineyard often contains a range of soil types. If it is sloped, it has higher and lower sections, and this has implications for temperatures and moisture. It may also have different aspects. Therefore, within just one vineyard, let alone one region, grapevines may be different. As a result, some people prefer to speak only of iconic or distinguished sites, those with outstanding *terroir*, rather than of larger areas such as regions. In McLaren Vale, for example, the Scarce Earth project tries to identify special sites.

This is not the approach I take. I prefer to look at *terroir* as a system of layers (see also Andrew Jefford's blog, 'Regionality and its myths'): Shiraz from Australia is different from Shiraz from France. At the next level, Shiraz from the Barossa is different from Shiraz from Victoria. Then, and this is where sub-regionality comes in, Shiraz from one part of the Barossa is different from Shiraz from other parts. And finally we have specific sites which create very specific and unique wines – Hill of Grace in Eden Valley would be a good example.

Another way of putting this is that *terroir* differences between major regions are predominantly influenced by climate or topography. Within a region, where climate does not vary too much, soil differences may account for the majority of differences in grape flavour and structure. This may then be amplified further on specific sites.

Terroir in the Barossa

The discussion so far has shown that the concept of *terroir* is not a very fixed and defined one. What then is important for the Barossa Zone? After studying the region and talking to many winemakers and viticulturists, I find three factors are most significant:

Temperature and temperature range

Temperature and temperature range are critical for the grape-ripening process. Factors which affect temperature are elevation, wind, latitude and soil. While one might at first think that there is not much difference across the Barossa, closer inspection shows significant differences: elevation for vineyards varies from 180 m to 400 m, and gully winds have a strong influence in some areas. Comparisons with Europe show that such differences can be significant for *terroir* effects. In fact, harvesting time for Shiraz in the Barossa ranges over four weeks every year. I will pay special attention to elevation for its influence on temperature.

Available moisture (rainfall and water-holding capacity)

Moisture is the amount of rainfall and how well it is held. Rainfall in the Barossa varies from 450 mm to 650 mm per year. The water-holding capacity of the soil

Some representative Barossa soil types. Clockwise from top left: Scholz Vineyard, 30 cm loamy top soil, 40 cm clay over calcrete; Kaesler Old Vine Shiraz Vineyard, red brown earth, 35 cm loamy sand top soil, 30 cm clay over calcrete; Hentley Farm, The Beast Vineyard, 30 cm loamy top soil, 40 cm red clay over hard rock; Schrapel Vineyard, 50 cm sandy top soil over 70 cm clay.

depends on its composition – the amount of sand or clay in it, for instance. This capacity can vary significantly within the Barossa.

Soil

The Barossa region is very old, much older than many other wine regions of the world, and its soil has been shaped over millions of years. The result is a lot of variety and complexity. Not surprisingly, the soil varies significantly across the Zone. The map below shows the major soil types across the region.

Barossa Region

Simplified soil map

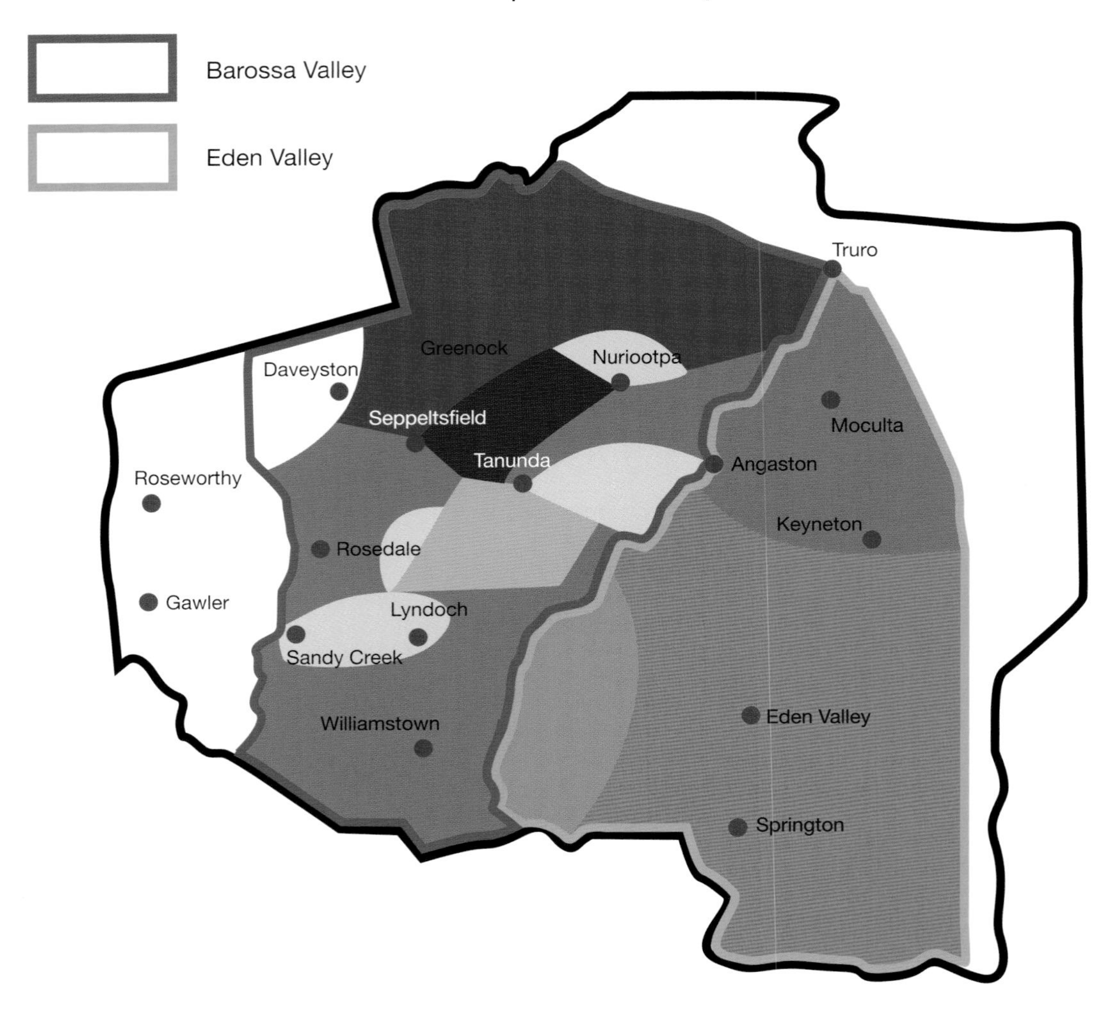
Barossa Valley
Eden Valley
Truro
Greenock
Nuriootpa
Daveyston
Seppeltsfield
Moculta
Tanunda
Angaston
Roseworthy
Keyneton
Rosedale
Gawler
Lyndoch
Sandy Creek
Eden Valley
Williamstown
Springton

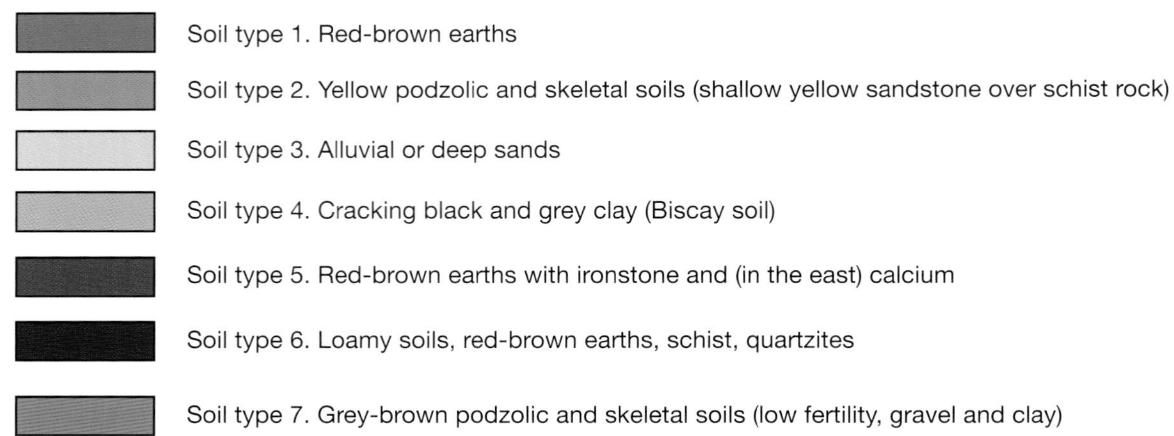
Soil type 1. Red-brown earths
Soil type 2. Yellow podzolic and skeletal soils (shallow yellow sandstone over schist rock)
Soil type 3. Alluvial or deep sands
Soil type 4. Cracking black and grey clay (Biscay soil)
Soil type 5. Red-brown earths with ironstone and (in the east) calcium
Soil type 6. Loamy soils, red-brown earths, schist, quartzites
Soil type 7. Grey-brown podzolic and skeletal soils (low fertility, gravel and clay)

The dominant soil types are:

1. Yellow podzolic and skeletal soils (shallow yellow sandstone over schisty rock)
2. Grey–brown podzolic and skeletal soils (low fertility, gravel and clay)
3. Red–brown earths

3a. Red–brown earths with ironstone and (in the east) calcium

4. Alluvial sands

4a. Deep sands

4b. Shallow sand over clay

5. Cracking black and grey clay (Biscay soil)
6. Loamy soils, red–brown earth, schist, quartzites

These three factors are the key environmental factors in the Barossa. Other factors, such as slope or direction, which are critical in some regions, such as the Moselle region, are less important for a discussion of sub-regions in the Barossa. They may, however, be important for individual vineyards.

Methodology

The methodology I have used to define the sub-regions is based on a combination of approaches. It first involved an analysis of soil, elevation and rainfall data. Second, I interviewed winemakers and viticulturists to obtain expert views on the subject. Third, I was fortunate to be able to taste, at a number of wineries, Shiraz from different vineyards (mostly from different growers) but of a particular year. Some of these tastings were bottled wines from 2005 to 2007, but most were barrel samples of the 2009 vintage. My personal impressions are based on these tastings. And finally, some man-made geography played a role, for practical reasons: for example, the Sturt Highway forms the northern border of the Upper Central Flats. Borders of municipalities have not influenced my judgement, although place names have been used to label a number of sub-regions.

In this context, it may be worthwhile to point out that Bordeaux classifications are not based on *terroir* alone, but also include property considerations. For example, a château's classification will get expanded if it acquires a neighbouring property (which does happen from time to time).

A further significant question is the issue of boundaries. Some argue that it is either too early to finalise boundaries or that the boundaries would ultimately be arbitrary. This group prefers to talk about the core of an area; as a consequence, the area's characteristics would be presumed to peter out towards the boundaries. While I have some sympathy for this approach, I cannot support it. The concept of sub-regions is based on *terroir.* Logically, this requires geographical

Officials set up wines for a tasting of regional Barossa wines.

boundaries. I have decided to draw specific boundaries. In one or two cases, the grey zone is so significant that it needed to be recognised. There is no ultimate right or wrong here. In the end, wineries may want to sit down and agree on the boundaries, as they have done in the Napa Valley, for example. I believe that a move to specifically defined regions is inevitable. Let this be the first attempt.

Before I analyse how the above-mentioned factors influence the grapevines in different parts of the Barossa, I would like to briefly look at the historical development of winemaking there and the economic models used. This will help put the Barossa into the context of the world of wine.

The influence of history on Barossa winemaking today

2

Making premium table wine is a relatively recent activity in the Barossa despite the existence of some remarkable old vineyards. A number of key events are responsible for this situation.

The Barossa Valley was originally settled in the mid 1800s. Its fertile land received attention from English and German migrants. Ann Jacob is now thought to have been the first person to buy land with the objective of establishing a vineyard there. She purchased land at Rowland Flat, at the junction of Jacobs Creek and the North Para River, in October 1839. The property was called Moorooroo, meaning 'big waterhole'. She and her brothers established a mixed farming business, including a vineyard.

Most settlers came from Germany, which was experiencing increasing unrest during the 1840s. The migration was the result of a deal between George Fife Angas, a grazier who owned pretty much all the Barossa, and Lutheran Pastor Kavel in London, who wanted to bring out Germans so that they could practise their religion more freely. They brought an approach to the Barossa which is still alive today: they were frugal, had a strong work ethic and a commitment to the church. Wheat, wool and wine were the focus of agriculture in the valley, initially. The early farms were chosen with an eye to the availability of water, so most were established in the river flats, on alluvial soil. The oldest vineyard still in operation is probably the 'house block' of the Langmeil winery, next to the North Para River – it was established in the 1840s.

These Lutheran settlers mostly became small landowners in the valley and established Bethany as the first village in 1842. The wealthier English settlers, in contrast, often established larger sheep and cattle grazing properties in the hills. Overall, mixed farming was the most prevalent approach, as it minimised risk, and horticulture and viticulture became the focus. The Germans had a long association with wine – in fact, wine had been made in Silesia since the 12th century – and it was part of their approach to self-sufficiency. Interestingly, the Germans saw themselves as caretakers of the land, and their aim was to leave it in better shape for their children. This explains the many significant vine-growing families who are now in their fifth or sixth generation of growers. It could be said that valuing *terroir* was implicit in their philosophy.

These beginnings helped determine the direction of winemaking in the Barossa. The starting point was growing grapes for survival: that is, for trading

Church at Bethany

and selling. Grape quality was not a major consideration. This was not the case in other winemaking regions. The region of Bordeaux, for instance, in contrast, is often characterised as an area where millionaires make wine for millionaires. The vastly different historical and economic and cultural circumstances in these two areas have led to different winemaking principles being adopted in them.

Domestic wine consumption began with the English population in Australia, who had enjoyed drinking French or German wine with meals in Europe. In 1858, the introduction of the Wark Act allowed the number of licences for distilleries to increase significantly. The winegrowing area in the Barossa increased almost tenfold in the subsequent decade as a result, because in those days winemaking depended on the use of spirits for preservation. As import duties were reduced in England during the 1860s, wine exports to the United Kingdom increased. The Pewsey Vale winery, in particular, achieved some recognition and success.

The second major factor in explaining today's industry was the repeal of the Wark Act in 1877. The subsequent reduction in the number of available distillery licences meant that many winemakers left the industry altogether or focused on grapegrowing rather than winemaking. As a result, the industry split into two distinctive parts: a larger number of grapegrowers and a few companies which made and marketed wine. The largest of the latter were Seppelt, Thomas Hardy & Sons, Chateau Tanunda, Yalumba, Saltram and Kalimna. This specialisation became an important feature of the Barossa (and the Australian) wine industry. The result, initially, was a grower focus on yield, and a marketing and branding focus by the wine companies.

The stars of the Barossa were fortified wines, such as port and sherry. The 'Imperial Preference' of England, which allowed reciprocal tariff-free trade, advantaged wine imports from the colonies. Exports to Britain were crucial for the Barossa, which had established itself as an important part of Australia's wine

Graveyard at Bethany

industry. The Barossa accounted for 25 per cent of Australia's production before the Great Depression of 1929. In that year, the Barossa suddenly faced significant overproduction, the result of a large vintage plus exports to Britain dropping by more than 50 per cent. Between the world wars, the industry went through cycles of growth and decline.

The third major factor influencing winemaking in the Barossa today occurred after World War II: red table wine began to be made. In the 1950s, Colin Gramp of Orlando started to produce table wines in volume. Barossa Pearl, the 'wine of the people', was his biggest success.

The beginnings of Grange Hermitage in 1951, by Max Schubert of Penfolds, are well documented. The winemaking techniques adopted by Schubert became the model for premium winemaking in the Barossa for the next 50 years. Schubert's mission, on his visit to Europe in 1949, was to learn about sherry production, but a side trip to Bordeaux and a tasting of 40- to 50-year-old wines led to his ambition to create a great Australian table wine, a wine that would improve for a couple of decades and live for a couple more. As Cabernet Sauvignon was in short supply and the other Bordeaux varieties were hardly available at all, Schubert decided to base this wine on Shiraz. Initially, the grapes came from the Magill vineyard and another vineyard south of Adelaide. But the Barossa took the lead in later years. His objective was to create a big, full-bodied wine with maximum extraction of all the components in the grape material used.

The empirical approach to achieving acid and sugar balance and controlling fermentation that he used was based on research work done by Dr Ray Beckwith. There were several key aspects of this approach:

- head-down/submerged cap fermentation;
- use of untreated American oak hogsheads; and
- pressing before completion of primary fermentation and transferring direct to barrel.

This last decision in particular increased the volume of bouquet and flavour. Through the oak treatment, the fruit flavours had become bigger and more pronounced. Over time, this wine came to be regarded as the greatest wine in the southern hemisphere, so the use of American oak and barrel fermentation have become traditional aspects of Barossa Shiraz winemaking.

In this context, it is also necessary to point out the influence of major oak manufacturers. A.P. John Coopers, for example, started its business in the late 19th century at Chateau Tanunda and is today managed by the fourth generation, from an independent site in Tanunda. Its toasty American oak contributed significantly to the flavour profile of most Barossa Shirazes made until the 1990s; a wider range of oak treatment has been employed since then.

These types of wines could also be successfully paired with Barossa food. The Barossa has a long and proud history of food culture, going back to early German settlement. Its influence mixed with English culture and local produce. The food is rich, and often heavy. Winter sees stews, dumplings, sauerkraut and broad beans. Summer sees picnic baskets full of wursts, sausages, ham, bacon and cheeses. Game such as hare and goose is popular. These foods require big, rich wine styles.

Red wine's popularity – and its production – started to increase during the mid 1960s. From 1964/65 to 1969/70, red wine sales increased by over 150 per cent, and Barossa wines were the stars. Penfolds won three Jimmy Watson trophies for its predominantly Barossa wines between 1964 and 1968, and

Saltram won one. The Jimmy Watson trophy, given for the 'best 1-year-old dry red table wine', has long been the most keenly sought wine award in Australia.

Many family wineries could not fund the growth they needed to remain successful so they sold their businesses to larger – and in some cases overseas – companies. Those buyers were mainly companies that bought and sold a range of consumer goods, rather than just wine, so branding became an even stronger focus during this time. Grapegrowing was seen as an input cost, so quite a lot of it moved to lower-cost irrigated regions, thus reducing the price at which these larger producers could sell their wine. In the Barossa itself, many growers responded to this by introducing mechanical harvesting (to lower the cost of production) and irrigation (to increase yields).

Other growers continued to focus on quality. They survived because their grapes were put into high-quality wines which could be sold at premium prices. Again, Barossa wines won the Jimmy Watson Trophy repeatedly: six times during the 1970s. These pressures on price and production were replicated in other farm commodities, so while some small vineyards survived, most small dairies and other mixed farm activities did not.

Cask wine was introduced in the mid 1970s, and table wine took over from fortifieds, which had still dominated wine sales in the 1950s and 1960s. Partly in response to the lower quality mass-produced red wines, many consumers shifted to white wine. In 1977 Saltram announced that it would not buy any red grapes from its growers the following year; it had lost much of its market as a result of this shift in consumer preferences. Interest in red wine continued to decline, and some vines were eventually pulled, beginning in 1982. The South Australian Government Vine Pull Scheme was launched in 1986. Under this scheme, growers who removed old vines and left land unplanted were reimbursed. As a result, many old Shiraz, Grenache and Mourvèdre vineyards were lost.

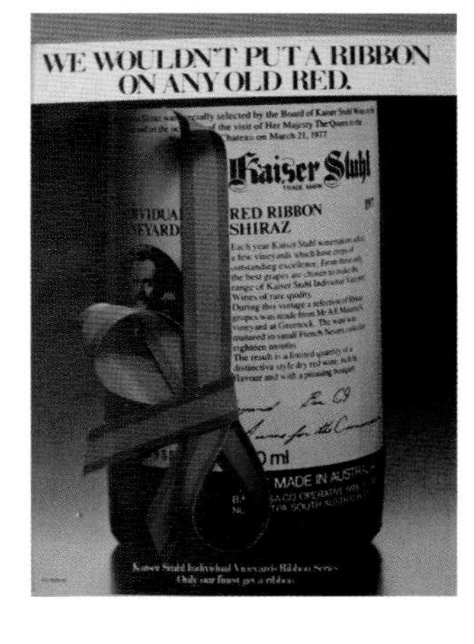

The modern Barossa grew out of this low point. The way this unfolded is the fourth major factor in explaining the current state of the industry. Peter Lehmann, Saltram's winemaker in 1977, decided to honour the growers' contracts, and bought the 1978 fruit himself (with a little financial help from his friends). A year later, he founded the Masterson Barossa Vignerons winery, in time for the 1980 vintage. The winery, which later became Peter Lehmann Wines, faced many obstacles, but its commitment to its growers and the Barossa never waned.

Three new wineries, all focusing on old vine Shiraz, were started in 1984: Rockford, Charles Melton and Heritage. These wineries paid growers above market rates, which at the time were less for Shiraz than for Cabernet Sauvignon, Chardonnay, Riesling or Semillon. Major icon Shirazes which celebrated the Barossa's old vine heritage were launched during this time: St Hallett's Old

Block, Rockford's Basket Press, Peter Lehmann's Stonewell, Charlie Melton's Nine Popes and Grant Burge's Meshach. Only a couple of years later, prices rose significantly. In 1988, Shiraz grapes achieved $800 per tonne – they had been $275 in 1985. 'Brand Barossa' started to be recognised as a red table wine brand, and Shiraz in particular was standing out.

The 1990s saw the revival of the Barossa. International recognition increased dramatically, and Barossa wines and winemakers won many trophies. The high point came in December 1995, when the 1990 Penfolds Grange was named Wine of the Year by *Wine Spectator* magazine. It can be argued, however, that these rewards had their downside, and that the extensive Australian Wine Show system that had by then developed in fact led to an increasing homogenisation of Australian wine, as a result of its technical and prescriptive approach.

However, in the 1990s, Robert Parker, the US wine critic, suddenly found big, ripe Australian wine, in particular Barossa Shiraz, to his liking. Chris Ringland's Three Rivers Barossa Shiraz was rated 99 points for the 1993 and 1995 vintages, and 100 points for the 1996. Greenock Creek's Roennfeldt Road Shiraz was awarded 100 points for three vintages between 1995 and 1998. Torbreck's RunRig and Rolf Binder's Hanisch both scored 99 points during this time. These wines became cult wines overnight, with prices to match. This was initially a healthy development for Australian wine, and for the Barossa in particular. Exports and prices, across the board, increased dramatically. Many new boutique wineries started up. However, after 2001, with the end of the dot.com boom, export prices for Australian wine started to fall, as the number of speculative buyers declined; the Global Financial Crisis of 2008/09 also resulted in a significant reduction in export volumes and prices.

The positive aspect of all this development was the creation of some very different wine styles. This is the fifth major factor contributing to today's situation. So-called Robert Parker wines were big, alcoholic, in your face – very different from the more restrained European style favoured at Australian Wine Shows. This divergence led to a famous clash of words between Robert Parker and James Halliday, Australia's best known wine writer, in 2005. James Halliday, in a speech at the Wine Press Club of New South Wales, argued that Robert Parker's Australian wine recommendations favouring 'monstrous red wines' were at odds with Australian Wine Show results, and Parker responded by accusing James Halliday of being a 'Euro-imitator' and favouring wines made to a bland format: 'add acidity and then some more acidity to eliminate any trace of a wine's place of origin'. Importantly, this debate made obvious how different Shirazes could be, even Shirazes from the same region.

Winemaking in the Barossa has since become more self-assured. Different winemaking techniques and approaches to oak management are now used to show different qualities of Shiraz, depending on winemakers' preferences and where the grapes are grown. Today, Shiraz is much more than 'brand Barossa', and the search is on to identify which areas are best suited to what kind of Shiraz.

Barossa wine is still not 'made by millionaires for millionaires', as in Bordeaux, but perhaps it can be described as 'character wine', made by Barossa characters.

Having some understanding of the history of winemaking in the Barossa is important in order to understand why it, as a premium wine region, is different from Bordeaux and Burgundy, which I will use as reference points from time to time. That history includes traditions of who owned land, who made wine, how wine was sold, and more.

In the Bordeaux region, for instance, the tradition since the late 17th century has been for vines to be grown on land surrounding chateaus. The grapes, and the wine, were thus owned by the producer, who would sell the finished wine to a merchant based in Bordeaux, who would then on-sell the wine, much of it to London and northern Europe. It was important that the wine could be associated with the particular 'chateau'. A focus on lower yields and *terroir* was beneficial to the producer/winemaker, as it improved quality, and made the wine rarer (as less was produced because of the lower yields). The first of these winemakers were wealthy magistrates. Later, in the 19th and 20th centuries, bankers and industrialists who could afford significant investments, for example in storage facilities and new barrels, bought the chateaus and the winemaking businesses that went along with them.

In Burgundy, however, the dominant force for centuries has been the individual peasant grower, most of them with very small plots as a result of the division of properties among heirs to an estate. They sold their grapes to merchants or negociants who made and sold the wine. In the last two decades there has been a change, and more and more growers have decided to make their own wine. Some have only very small vineyards, and have decided to take on the role of micro-negociants themselves, and to then compete with the larger growers. So in contrast to Bordeaux, the Burgundy wine culture could be defined as rural.

In the Barossa, too, grapegrowers were divorced from winemaking in the 19th century. As a result, their main interest was to increase yield in order to maximise volume and therefore revenue. Wine companies wanted to buy grapes cheaply. They focused on their wine brands, because consistency was key. Virtually all wine was blended – between growers, and often between regions.

The picture today is much more diverse.

The Australian wine industry is dominated by two major companies, Treasury Wine Estates (formerly Foster's) and Constellation Wines Australia, both with a focus on branding, scale, and blending across regions. Both own vineyards, and buy significant amounts of grapes from individual growers. However, some of their sub-brands take somewhat subtler approaches. Penfolds, for example, a brand steeped in Barossa history, are increasingly making the Barossa origin clear on their wines.

Many medium-sized companies, such as St Halletts, Rockford, Yalumba and Torbreck, source fruit from a stable of growers under long-term arrangements. Some own some vineyards themselves. These companies have a strong commitment to the Barossa and they too are increasingly valuing sub-regionality.

Some growers have gone into wine production themselves, reducing the amounts of grapes they are contracted to sell. Examples would be Schubert, Smallfry and Pindarie.

Some of these growers have emerged as estate-type concerns with more substantial vineyard holdings in one area. These include Greenock Creek and Kalleske.

There are also some small entrepreneurial – or virtual – winemakers. These are businesses that try to limit their investment in vineyards and sometimes only invest in their winemaking skills. Larger ones include Teusner and First Drop Wines.

These different approaches are creating a dynamic environment for premium wine, which may benefit the Barossa. The medium-sized companies place a strong emphasis on quality and they provide incentives to growers to produce high-quality grapes. Growers turned winemakers realise that without having scale, they need to sell wine at fairly high prices to survive. The same is true for the entrepreneurial winemakers.

The industry has come a long way from its origins – producing table grapes for survival. Within the last 20 years, larger companies have introduced systems and processes that allow them to produce high-quality wine even in large volumes. Many smaller producers with a focus on quality have managed to compete with the big-volume production companies.

The current economic conditions deliver major challenges to the wine industry. Unfortunately, many wine companies have responded by jumping on the latest trendy bandwagon and applying spin which ultimately undermines the credibility of the whole industry, such as:

- quoting vine ages that cannot be substantiated;
- promoting old vineyards, although only a couple of rows are really old;
- emphasising the intent to go biodynamic, while using chemicals;
- propagating the latest vintage as always the best or talking about how they brought in the grapes just before the heatwave struck or, in an undisputedly bad vintage, how the non-production of the premium wine will improve the quality of the standard wine;

- marketing the use of the basket press, although only a small part of the production goes through it

... and the list goes on.

The competition between producers has increased, and the need for differentiation is greater than ever before. The skills needed to do this successfully are also greater than ever before. The focus on *terroir* can provide a successful path, but it means accepting vintage variation and honesty in marketing. The starting point for being economically successful long term, I suggest, will be recognising the sub-regions of the Barossa.

We are a long way from a *grand cru* and *premier cru* classification (a classification of French vineyards reflecting their reputation for producing quality wine), and many winemakers I have spoken to are opposed to using such an approach, given the egalitarian ways of Australia. They also cite commercial difficulties: as many grapes are sourced from independent growers, any official acknowledgement of their vineyard quality could lead to a bidding war for their fruit and an increase in grape prices. However, the market works in this way anyhow.

In my view, what the combination of consumer interest and economic necessity will produce is an irreversible momentum towards a focus on unique wines, and that means a focus on *terroir*.

3

Approach to sub-regions

The Barossa Valley Region is not a large geographic area. It is an hour's drive north of Adelaide and extends approximately 32 kilometres from Williamstown in the south to Koonunga in the north, at latitude 34°S, and 15 kilometres east/west from the Eastern Slopes to the Region's border in the west. The Barossa is a hot region in the context of winemaking, with a fairly wide average temperature range – higher than the average temperatures of Bordeaux and even of the Douro Valley in Portugal.

Relationship of monthly average temperatures (°C)

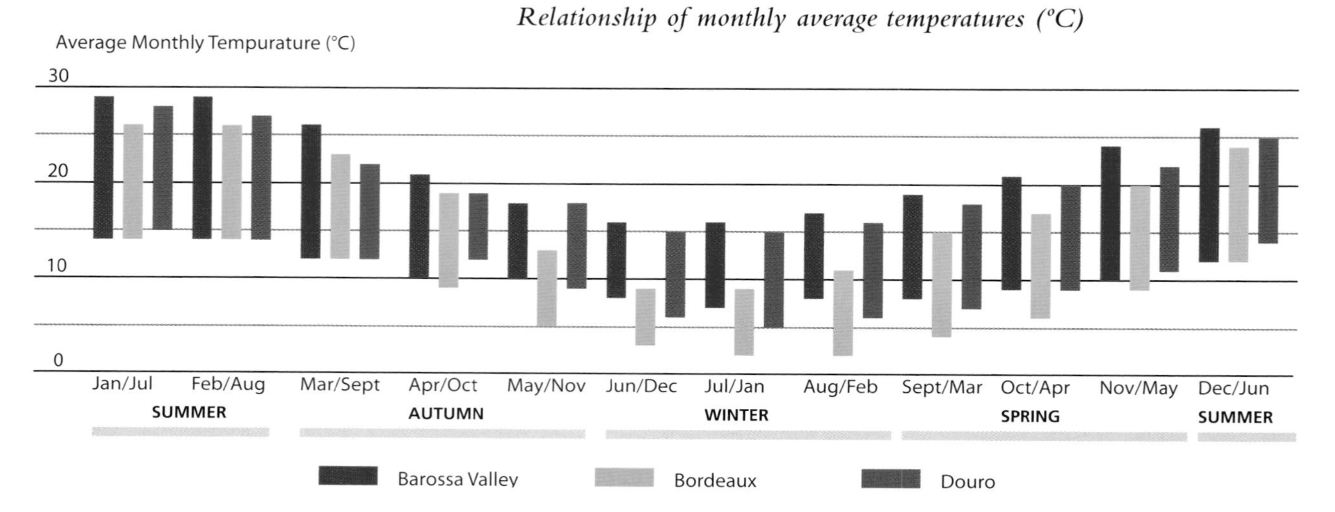

In relation to the major criteria relevant for the Barossa, there are significant differences within the region. The elevation varies from 178 m at Lyndoch to 400 m at Moppa and the Eastern Slopes.

The soil, too, is incredibly varied, due to the age of the land. Many other New World wine areas – in New Zealand and South America, for example – have developed on sediments less than 12,000 years old. The vineyards of Bordeaux are grown on terraces 1 million years old. The Central Valley of the Barossa Region began to form 35 million years ago, and many different land surfaces have developed since then.

The average annual rainfall for Nuriootpa is 470 mm, but rainfall during the hot summer months tends to be low, and evaporation exceeds rainfall between October/November and April. Chris Ringland, a winemaker with experience in both the Barossa and Spain, notes that there is significantly higher evaporation in the Barossa than in Europe during the summer months. This means that vines are more exposed to dry soil in an average year.

Mean rainfall at Nuriootpa (mm)

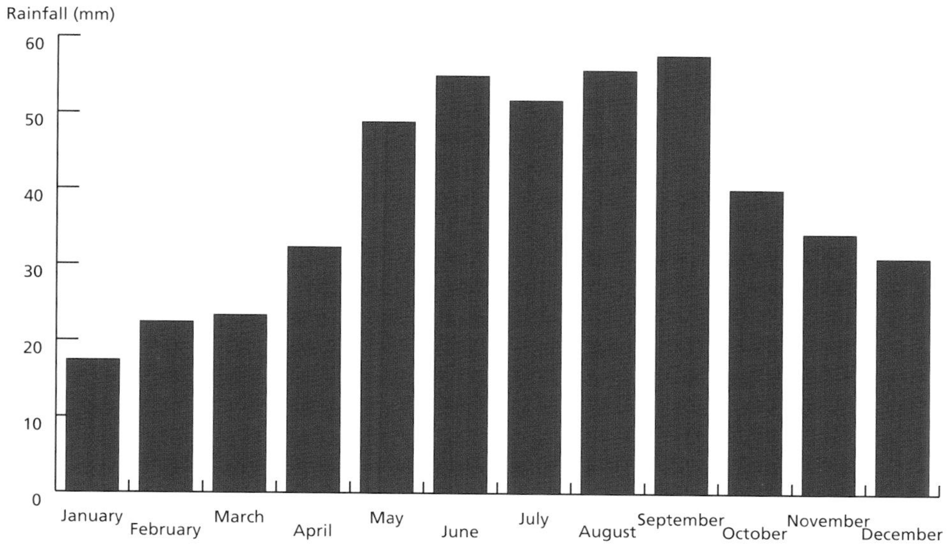

Furthermore, while detailed rainfall data is not available for different areas, it is clear that it differs as well. The average rainfall in Williamstown, for instance, tends to be more than 700 mm per year, whereas it is only 450 mm in the western part of the Barossa.

The major towns are Tanunda (population: 5000), which is the historical centre, Nuriootpa (similar in population), which is the more modern administrative centre, and Angaston (population: 2000), in the hills. Other towns that have a population of less than 1000 people but that are nevertheless relevant to wine are Williamstown, Lyndoch, Rowland Flat, Bethany, Marananga and Greenock.

In reviewing the *terroir* data, as well as interviewing many winemakers from the Barossa, it becomes clear that the sub-regions of the Barossa can be grouped into four sections. I call them:

- Central Valley;
- Southern Valley;
- Western Ridge; and
- Northern Barossa.

The Central Valley is the heart of the Barossa. This is where most of the townships are located. It is formed by the Para River, and characterised by alluvial soils. The oldest vineyards are found in the Central Valley.

The Southern Valley could be seen as the southern extension of the Central Valley, but it has different climatic and soil conditions and is therefore regarded as a separate section.

The Western Ridge is the undulating area arising from the Central Valley towards the west. The Seppeltsfield winery is its historic centrepiece.

Autumn in the Barossa
[Photo: Dragan Radocaj]

The Northern Barossa is the area north of the Sturt Highway, undulating in the west, flat in the eastern part. This area has many vineyards, but not many wineries, and is therefore less frequented by tourists.

I have identified eleven sub-regions of the Barossa Valley Region. I have labelled them as place names. A place name is an important descriptor of the history and identity of the area. Some wineries claim trademark rights to such names to the exclusion of others. This practice needs to stop. A place name, with all its connotations, should not become a brand.

Interestingly, the tannin structure of Shiraz appears to be a more consistent predictor of the sub-regions than flavours are. The grouping into the four sections is shown in the table below. A peculiarity occurs in the Central Valley. Some areas within it show quite distinctive characteristics, but they do not neatly border each other. I will show and describe these as Light Pass, Vine Vale and Bethany.

Sub-regions of the Barossa Valley Region

Central Valley	Southern Valley
Lower Central Flats Upper Central Flats Eastern Slopes	Lyndoch Williamstown
Western Ridge	**Northern Barossa**
Gomersal Marananga/Seppeltsfield Greenock	Moppa Koonunga/Ebenezer Kalimna

The sub-regions are shown on the following map (over page).

The following sections of the book will describe each sub-region, its *terroir* characteristics, and the influences on Shiraz wine from it. They will also show major labels of premium wine sourced from the sub-region and list interesting wineries in the area.

BAROSSA and EDEN VALLEYS

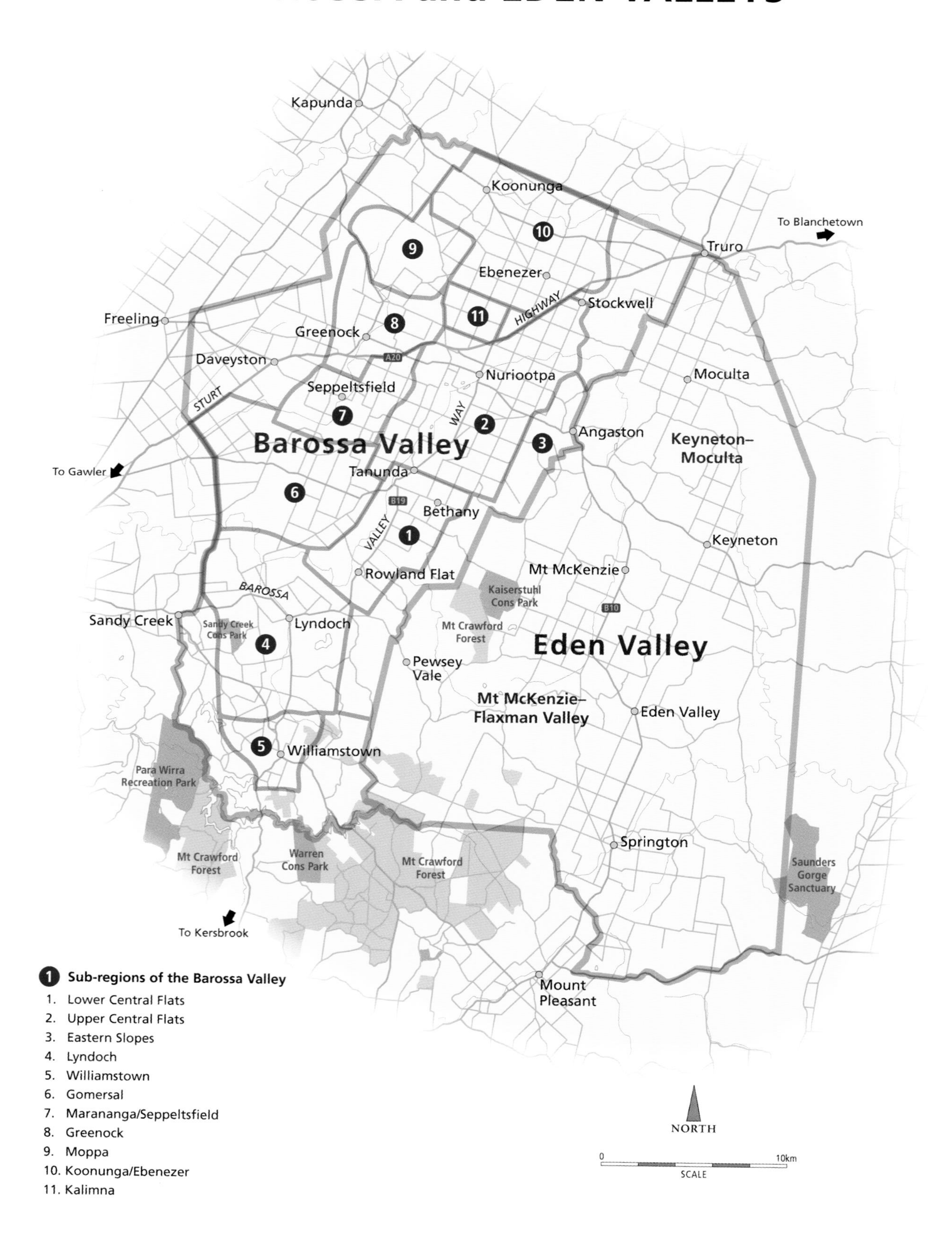

Kapunda
Koonunga
To Blanchetown
Truro
Ebenezer
Stockwell
HIGHWAY
Freeling
Greenock
Daveyston
A20
Nuriootpa
Moculta
Seppeltsfield
STURT
WAY
Angaston
Barossa Valley
Keyneton–
Moculta
To Gawler
Tanunda
B19
Bethany
VALLEY
Keyneton
Rowland Flat
Mt McKenzie
BAROSSA
Kaiserstuhl
Cons Park
B10
Sandy Creek
Sandy Creek
Cons Park
Lyndoch
Mt Crawford
Forest
Eden Valley
Pewsey
Vale
Mt McKenzie–
Flaxman Valley
Eden Valley
Williamstown
Para Wirra
Recreation Park
Springton
Mt Crawford
Forest
Warren
Cons Park
Mt Crawford
Forest
Saunders
Gorge
Sanctuary
To Kersbrook
Mount
Pleasant
Sub-regions of the Barossa Valley
1. Lower Central Flats
2. Upper Central Flats
3. Eastern Slopes
4. Lyndoch
5. Williamstown
6. Gomersal
7. Marananga/Seppeltsfield
8. Greenock
9. Moppa
10. Koonunga/Ebenezer
11. Kalimna
NORTH
0
10km
SCALE

The Central Valley

4

The Central Valley is the heart of the Barossa. It has mainly been shaped by the North Para River and its alluvial soils of sand, gravel and clay. This is where the major towns – Tanunda, Nuriootpa and Angaston – are situated. Tourism is focused on this part of the Barossa, and the major restaurants and hotels/motels are in this area. Many famous wineries have their cellar doors in the Central Valley and a number of old vineyards are located here.

This section consists of three sub-regions. The Upper Central Flats are different from the Lower Central Flats, mainly topographically. The Eastern Slopes is a separate sub-region because it is at a higher altitude and has different soil and wind conditions.

The Lower Central Flats

I have defined the border between the Lower Central Flats and the Upper Central Flats as Basedow Road, which leads from Tanunda to the Barossa Ranges. North of here, the land is quite flat, whereas rolling hills dominate to the south. Basedow Road is also where the black and brown black soils finish in the north.

The Lower Central Flats

This border means that the township of Tanunda is basically split between the two areas, but I have allocated it to the south because of its close relationship with Bethany and Langmeil.

Tanunda is the most central town in the Barossa. It was established in 1848. It sits at an elevation of 260 m, and its average rainfall is about 550 mm. The population of Tanunda is close to 5000. The town is still dominated by its seven churches, but there are also a number of cafés, hotels, restaurants and sport and leisure facilities. The Barossa Wine and Visitors Centre is located in its main street.

It all began 2 kilometres from Tanunda, in Bethany. Here, about 100 Lutheran migrants established the first village in the Barossa in 1842. Houses were built along the village street, with farming land stretching out far into the distance. However, Bethany never became a business centre. The main road bypassed it and ran through Tanunda, which also became the official postal centre for the area.

Names can be confusing in the Barossa. Langmeil, west of Bethany, was the second village, established one year later. It was renamed Bilyara during World War I, but reverted to Langmeil in 1975. However, Tanunda developed more quickly, and as the villages were joined, they became Tanunda.

Smaller settlements in this region are Krondorf and Rowland Flat.

Terroir

The elevation of the Lower Central Flats does not vary a lot. It changes from 230 m in the south to 260 m in the north. As mentioned, this area is characterised by rolling hills and numerous creeks. Essentially, the surface soil composition is similar to that of the Upper Central Flats, but the rivers, such as Jacobs Creek, have eroded the landscape more significantly and washed the sediments into the Gawler River and Gulf of St Vincent via the North Para River. The heavier erosion explains the hillier nature of the region.

There are pockets of difference. The flatter parts north of Krondorf Road have been less affected by erosion. From here to Bethany, the soils are black and brown–black (called Biscay soils), and heavy-textured. They extend as far as the northern border of this sub-region. The subsoils are impermeable clay. Waterlogging is a problem in wetter years, as it leads to too much vigour, and in dry years cracking earth is a problem. Near the Eastern Slopes, these soils mix with red–brown earth.

Waterlogging can also be a problem in the creek areas, where the groundwater table rises into the top 10 cm of the alluvial soils.

The other area which is distinctive is a small area south of Jacobs Creek. Below the loamy surface are subsoils of bright red colour, mixed with red granite pebbles. These sediments are landslide muds and pebbles which have come off higher ranges, probably as a result of earthquakes. They sit 5–10 m above river levels and create a very uneven surface.

Over 80 per cent of the rain in this area falls during winter. During the summer months, evaporation is greater than rainfall, making water management critical.

Major wineries

There are many famous cellar doors in the Lower Central Flats. The ones described here (and throughout the book) focus on fine wine, especially Shiraz. Although the cellar doors are in the Lower Central Flats, this does not mean that their wines are made from grapes from this area. As has been noted earlier, many wineries blend their wines, and many are supplied by growers – whose vineyards could be located anywhere in the valley. Jacob's Creek, Turkey Flat and Bethany do have major holdings in this area. (Wine labels shown in this book relate to wines made from grapes grown entirely or predominantly in the area under discussion.)

Chateau Tanunda is an iconic building and Australia's largest Chateau. It was constructed in 1890, from brick and bluestone quarried at Bethany. It has been fully restored, and has expansive gardens, a famous cricket pitch and an impressive cellar door. The first grapes grown near the Chateau were grown in the 1840s. Today, the Chateau includes a basket press winery. Grapes are sourced from the estate and from independent growers. The winery has access to a number of old vine Shiraz vineyards, including the Cricket Block at the Chateau.

Most Shiraz wines, including the flagship The Everest Chateau Cru Shiraz, are blends, but the new 'Terroirs of the Barossa' series intends to capture the specific flavours and structural profiles of the Northern Barossa, Greenock and Lyndoch, from individual vineyards that are over 40 years old (except Lyndoch, where the vineyard is 15 years old).

The Lyndoch Shiraz comes from a vineyard just north of Lyndoch. Its altitude is 185 m and its soil is brown earth, gravel and ironstone. This wine typically displays soft cherry fruit and dark chocolate. The Ebenezer District Shiraz comes from the Lowke family vineyard, which is at 280 m altitude, and has sandy loam and grey clay soil. This wine is typically full flavoured, with red fruit, leather and cedar characteristics and talc-like tannins. The Greenock Shiraz comes from the Matchoss vineyard west of Greenock, which is at an altitude of 290 m and has a soil profile of heavy brown earth and ironstone. The wine shows rich plum, licorice and dark chocolate flavours.

These wines are made in the same way, with small batch fermentation (started with cultured yeast) and gentle basket pressing used to preserve the unique

Old vines at Turkey Flat

qualities the respective vineyard has to offer. They are matured for 18 months in French and American oak and bottled unfiltered.

Turkey Flat is just across the road. The cellar door is a restored bluestone building – it was originally a butcher's shop, dating from 1860. Part of the Section One vineyard, near Tanunda Creek, dates back to 1847. The winery owns other vineyards in Bethany, Ebenezer and Stonewell. It has been in the hands of the Schulz family since 1865. They took the step from growing grapes to producing their own labelled wine in 1990. The Turkey Flat Shiraz takes in fruit from the old vines as well as newer plantings. Single-vineyard Shiraz wines from different sub-regions – Bethany, Ebenezer and Stonewell – are planned for release soon.

The Schrapels planted their first vineyard in 1852. ***Bethany Wines*** is still in the hands of this family today. The winery was built in 1981 in a quarry at the edge of the Barossa Ranges. Bethany Wines has significant plantings in this area, with vines ranging from 12 to 80 years old, and a number of different Shirazes are made from them, including the GR Reserve and the LE.

The ***Glaetzer Wines*** winery is in Tanunda, but it has no cellar door. Colin Glaetzer, after a successful career as the winemaker of Barossa Valley Estates and the creator of the E&E Black Pepper Shiraz, started his own label in the mid 1990s. Today his son Ben is the winemaker. He has been involved in many ventures, though he is still young, but his prime focus is the Glaetzer wines, with their striking Egyptian labels and names.

Ben's philosophy is to harvest according to grape flavour and use analysis as a backup only. Fermentation is run very cool, not exceeding 18°C, in order to preserve fresh fruit characteristics and to prevent unbalanced tannin extraction. His desire is to create wine with 'personality', reflecting the origin of the fruit. Softness, elegance and approachability are typical of the style.

His flagship Amon-Ra Shiraz is one of the most highly regarded from the Barossa. The grapes come from the low-yielding Hoffmann vineyard in the Koonunga/Ebenezer sub-region. Some of the vines are thought to be over 130 years old, but younger vines – a mere 50 years old – are also included. The grapes from this vineyard show great richness and concentration, and suit the desired wine profile well. Dark plum, blackcurrant and chocolate are typical flavours of the wine. The Bishop Shiraz comes from the same vineyards. It includes grapes from younger vines and is a little less concentrated. While the Amon-Ra uses almost 100 per cent new oak, less than 50 per cent new oak is applied to the Bishop.

The history of the iconic ***Rockford*** winery is described in the Robert O'Callaghan chapter of this book. The cellar door is on Krondorf Road, east of

the railway line, and is well worth a visit.

The winemaking philosophy is shaped by the way wines were made when Robert O'Callaghan was young. Rockford aims to preserve traditional Barossa wine styles, styles which are well established.

The winery equipment has been collected to preserve 'pre-industrial' winemaking in the Barossa. However, Rockford's higher volume wines are made elsewhere in modern facilities. The set-up of the Rockford winery resembles the way wine was made in the 1920s. The beautiful basket press is the best known piece of older equipment, but it is by no means the only example from this time. The buildings are made of Barossa ironstone.

The objective for its flagship wine, the Basket Press Shiraz, is for it to be a wine with a strong structure, but that is elegant and supple at the same time. Fruit comes mainly from Moppa and Eden Valley, but also from Marananga/Seppeltsfield and Greenock.

Rockford owns no vineyards, but has established long-term relationships and contracts with 40 or so growers.

St Hallett, by contrast, is a modern winery with plenty of stainless steel. It has contracts with more than 40 growers from all parts of the Barossa (and Eden Valley). There is enough room in the winery to keep up to 200 different Shiraz parcels. The winemaking is tailored to individual vineyards and does its best to encourage expression of place.

Three blended Shirazes are produced: Faith, Blackwell and the top of the range, Old Block. The fruit for Old Block is predominantly Lower Central Flats and Lyndoch, with 20–25 per cent Eden Valley fruit. In this wine, the power and richness of the Barossa is married to the elegance and complexity of the Eden Valley fruit. The flavour profile is aromatic, quite ripe, but not very tannic. The Blackwell fruit comes from Marananga/Seppeltsfield, Greenock and Ebenezer/Koonunga. This wine shows the more rich and robust style that is typical of the northern and western regions of the Barossa. As Toby Barlow, the winemaker, says: 'Ebenezer is the flesh and body, while Seppeltsfield provides the taut muscular skeleton of the wine. Greenock and Moppa add higher notes and purity.'

Across the road from Rockford is the ***Charles Melton*** cellar door. The winery produces three straight Shirazes, but its best known wine is a Shiraz/Grenache/Mourvèdre blend, Nine Popes. Charles Melton sources 40 per cent of its fruit from its own vineyards and 60 per cent from growers, most of whom are in the Lower Central Flats and Lyndoch.

Next door is ***Kabminye Wines***. The name is an Aboriginal word meaning 'morning star' and was given to Krondorf during World War II. (Krondorf has since changed its name back, clearly.) The cellar door is in a stunning modern building, which also houses the restaurant, which specialises in genuine recipes of the early German settlers. The winery has a special interest in sweeter wines.

Their premium Shiraz is called Hubert Barossa Valley Shiraz. It matures in new French oak hogsheads for up to three years. It is made mostly from fruit from the low-yielding Zerk vineyard at Gomersal, where the grapes tend to ripen very early.

Grant Burge Wines is unusual in that it has two cellar doors. The one on Krondorf Road focuses on white wine, and the restored and attractive main cellar door is off the main highway north of Rowland Flat. However, it has been reported that this cellar door will be sold to Orlando, owner of Jacob's Creek. Grant Burge has extensive vineyard holdings, with a focus in the Lyndoch area.

The flagship wine is the Meshach Shiraz. Most of its fruit comes from the 90-year-old vines of the Filsell vineyard. As a result, this wine has good fruit concentration, depth and intensity. It is highly regarded for its elegance and stylishness. It is only released after 20 months or so maturation – mainly in American oak hogsheads – and a further three years in bottle.

The Filsell vineyard is also the main source for the Filsell Shiraz, the Meshach's younger brother. This rich wine tends to have very dark colour, with sweet berry fruit and dark chocolate flavours. The winery also produces a number of other Shirazes at different price points.

Jacob's Creek, just south of Grant Burge, is – next to Penfolds and Wolf Blass – the best known winery in the Barossa. It is owned by a French company, Pernot-Ricard, and produces over 8 million cases per year. Its branding is so strong that in a UK survey, respondents named it the best known wine region(!) in Australia. The compound around the modern visitors' centre still includes the cottages of the Jacobs siblings. It is right next to these that the first vineyard was planted by Johann Gramp, in 1847.

Most of the Jacob's Creek wines are blended from many areas, but their flagship Barossa Shiraz, the Centenary Hill Barossa Valley Shiraz, comes mostly from vineyards along Jacobs Creek itself. The exact composition changes from year to year. The vines are up to 100 years old. Its name comes from a place where, in 1947, Fred Gramp planted a Moreton Bay fig tree on a hill overlooking the winery to commemorate 100 years since the first vines were planted. The individual fruit parcels are treated separately in the winery. Malolactic fermentation takes place in new and 1-year-old American hogsheads, where the wine is also matured for 21 months. After the final blend is created, the bottled wine will be held for another 18 months to ensure complete integration of oak and fruit. The completed wine tends to be soft and full-bodied, with rich and sweet dark fruit and berry flavours, supported by chocolate and spice.

Lou Miranda Estate established itself in the Barossa only in 2004, but it

Centenary Hill vineyard, Jacob's Creek

occupies the old Rovalley Wines site in Rowland Flat, and so has access to small plantings from as far back as 1898 and 1907. Their wide range of commercial wines includes the Old Vine Shiraz and the Cordon Cut Shiraz.

Tasting profile

What are the flavour characteristics of Shiraz in the Lower Central Flats? We need to remember that the *terroir* puts the basics in place. Vineyard management and winemaking techniques can either emphasise or alter these characteristics.

As discussed above, there are not many premium wines produced from fruit exclusively from the Lower Central Flats. I therefore had to taste a number of individual barrel samples from different wineries in order to determine the specific flavour and structure characteristics of this sub-region.

The vines tend to ripen early in these moderate conditions. The main feature of Shiraz from this area is that it is less astringent and shows less extraction than that from other parts of the Barossa. Fruit flavours vary from red berries to plum. Tannins stay very much in the background. Vigour control is very important in order to make expressive wine. Most Shiraz from this area is gentle and soft and shows evenly on the palate. Wine of lesser quality can be quite fruity.

The wines that come from the black soils around Bethany and Biscay Road are a little different. The vines ripen early there, but relatively slowly compared with other sub-regions, probably due to the gully winds in the afternoons and the cooler nights, particularly as the sites approach the Eastern Slopes. The fruit flavours here belong to the blue spectrum: plum, blackberry, blueberry. As in other parts of the Central Valley, the wines tend to be aromatic and lifted, and not overly tannic. These wines are not massive, but neither are they as linear as those from Eden Valley, for example. I also noticed more earthy flavours than in the other parts of this sub-region.

Vineyards

As discussed before, there are many creeks in the Lower Central Flats, and alluvial soils dominate. Many original vineyards survived here because there was enough water for the vines to last the distance.

The famous old vineyards in the Lower Central Flats include Schild's Moorooroo vineyard. It has only four rows left from the 1800s plantings, but the remaining Shiraz vines are still over 50 years old. Turkey Flat's Section One Shiraz plantings date back to 1847 and are among the oldest in the world: there are 11 rows, with a total of approximately 930 vines, left from this time. Jacob's Creek's Centenary Hill vineyard was planted in 1922 along the banks of Jacobs Creek.

Old vines

Lower Central Flats

BAROSSA VALLEY

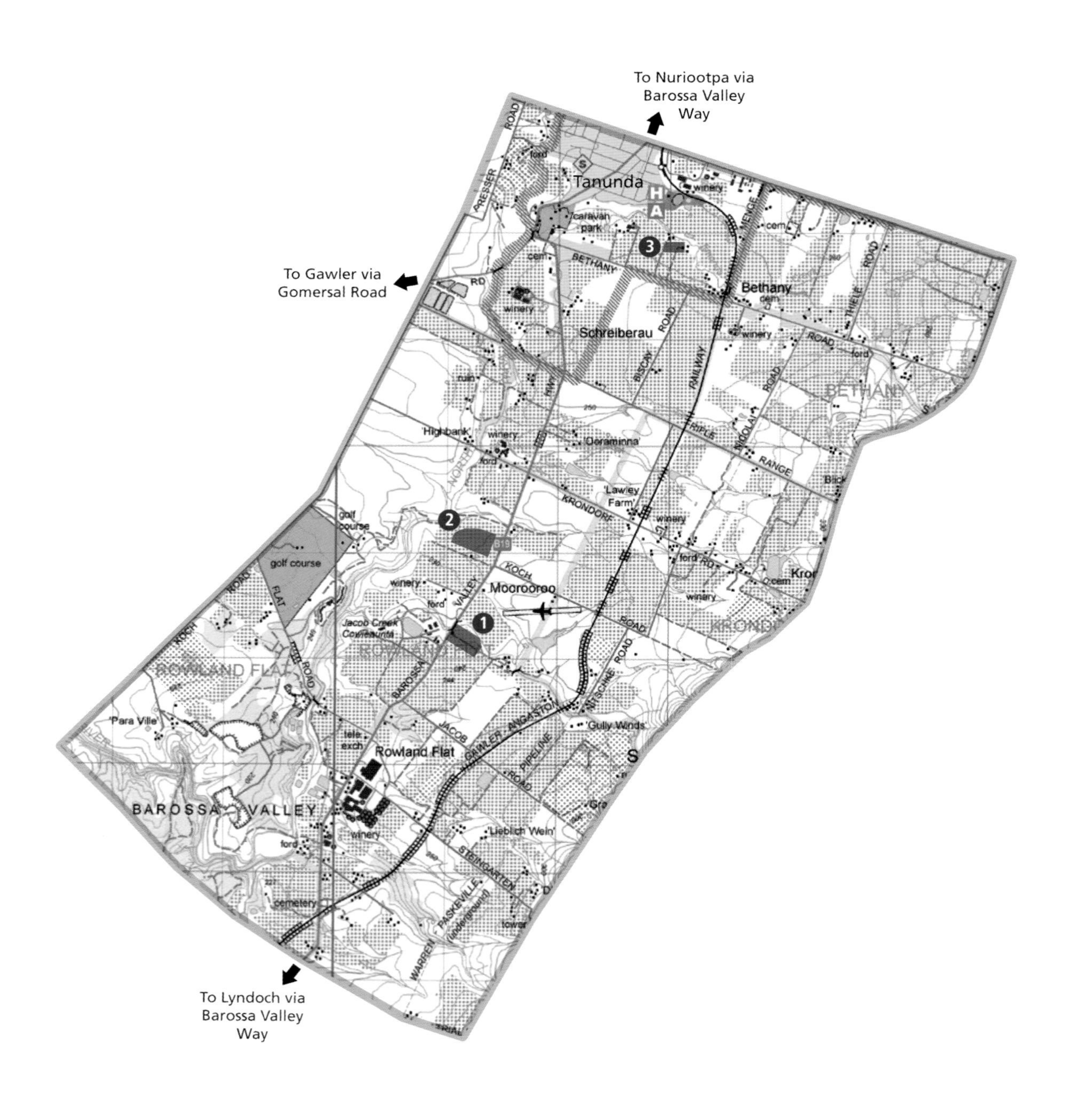

❶ Benchmark Vineyards

1. Jacob's Creek, Centenary Hill Vineyard
2. Schild, Moorooroo Vineyard
3. Turkey Flat, Section 1

0 2,500m
SCALE

A number of young winemakers who have ventured out on their own have expressed to me their frustration about the 'obsession' with old vines, as obviously 'old vines' cannot be achieved quickly. Old vines have deeper and more widely spread roots and are therefore better equipped to deal with weather variation, and thus more insulated from stress. Also, because they can draw moisture from the soil at a steadier rate, the grapes are more similar to each other and their fruit is often more elegant. Deeper soils are more heterogeneous than top soils. Therefore, vines with deeper root systems can express the specifics of the land more successfully. And finally, old vines have lost the vigour they had when they were first growing, so they carry small berries, which increases the skin to juice ratio, which in turn means greater fruit concentration. The end result is a lower yield and a more expressive wine. Grant Burge believes that the yields start to drop significantly when the vines get to 50 years old. This may occur earlier in other sub-regions, where the vines get stressed more often. The ability of old vines to express *terroir* is particularly important in the Lower Central Flats, where the soils are alluvial and more fertile and young vines have a tendency to produce high yields and diluted flavour. Because vine age is regarded as important, it is worth pointing out that the exact age of vines is often not proven and records are sketchy. Also, sometimes vineyards have a few old vines but others that are much younger.

The Upper Central Flats

The Upper Central Flats cover the area north of Tanunda up to the Sturt Highway. The western border is more difficult to define. Based on topography, I have included Stonewell, up to Stonewell Road, in this region. However, the Stonewell area is a real transition area between the Central Flats and Marananga/ Seppeltsfield. Also, it is quite small and does not really have a *terroir* of its own.

The centre of this region is Nuriootpa, or 'Nuri', as the locals call it, the commercial centre of the Barossa Valley. Nuri is a town of about 4500 people, with an elevation of 270 m and an average rainfall of 470 mm per year – slightly higher, and with 10 per cent less rainfall, than Tanunda.

The Upper Central Flats started to be settled soon after the establishment of Bethany, further south. Small 8-hectare mixed farms that grew wheat and had vineyards were established along the Para River and in Light Pass. Nuri started as a camping place on the way to the copper mines of Kapunda and Burra in the mid 1840s. The township was laid out in 1855. Wheat growing was the main activity, and a flour mill was erected in the town in 1874. At the same time, vineyards seemed to flourish, and vine cuttings from a number of different sources were brought into this sub-region.

In about 1890, phylloxera wiped out many vineyards in Victoria. In the following years, vine plantings in the Upper Central Flats increased significantly. The climate and the soil proved very suitable for vineyards, and they provided the best income for small farms. Today, pretty much all land in the Upper Central Flats that is not used for housing and community services is covered by vineyards.

Roses near Dorrien

Terroir

The Upper Central Flats are, as the name suggests, flat. The alluvial flats of the North Para River are quite narrow along the river bed from Tanunda to Nuriootpa. The sediments in the valley consist of sand mixed with gravel or clay and are quite young: less than 5 million years old. The sand and gravel is usually quite coarse, and can be up to 100 cm thick. The subsoil has a finer texture and tends to be quite deep as well. Sediment from the Eastern Slopes does not appear to have been washed onto the valley floor.

A number of areas can be defined a little more specifically. Vine Vale is the first. It is defined by alluvial sands, quite deep, on yellow and red clay. Vine Vale covers roughly the area between Basedow Road, Barossa Valley Way, Nuraip Road and Stockwell Road. The sand, combined with gully breezes at night, means that the temperature decreases more at night here than it does in other areas. The resulting broader temperature range leads to later ripening of the grapes and perhaps less fruit concentration and more red fruit, such as raspberry, and supple tannins. I found the flavour profile not consistent enough, though, to define Vine Vale as a separate sub-region.

The second area is Light Pass and Stockwell. The soil in this area, north of Angaston Road, is red–brown earth over limestone, and the area is windier than the area south of Tanunda. The gully winds in the afternoon keep the temperature from soaring and often prevent any shut-down of the grapes during the ripening process.

Stonewell, northwest of Tanunda, is a less fertile area, and grapes are more difficult to grow there. The soil is predominantly red–brown earth over white clay. As the vineyards are slightly sloping west to east, they bathe in the afternoon sun.

Major wineries

The main wineries with cellar doors in this sub-region, from south to north, are Langmeil, Peter Lehmann, Rolf Binder (Veritas), Whistler, Kaesler, Penfolds, Elderton, The Willows Vineyard, Gibson and Wolf Blass.

The ***Langmeil*** winery is steeped in history. The 1.5-hectare Freedom vineyard, planted in 1843, is probably the oldest still growing Shiraz in Australia and one of the oldest in the world. It was planted by Christian Auricht on the banks of the Para River. The Freedom Shiraz is still produced from its grapes as a single-vineyard wine. The first winery on this site (Paradale) was established in 1932. The business was sold to Bernkastel in 1972. After Bernkastel went into liquidation in the 1980s, the winery lay idle for some time before the current owners re-established it as Langmeil in 1996.

The winemaking philosophy is to ensure that the fruit shines through in the

Old Shiraz vines at Langmeil: possibly the oldest in the world.

final wine. The fruit is hand-picked. Key features of the process are minimal handling, open fermentation, basket pressing, and no fining. New and old French oak hogsheads and barriques are used.

Today, the second important vineyard is the adjacent Orphan Bank Vineyard. It consists of vines that are up to 140 years old – they have been replanted here. From it, a second single-vineyard Shiraz, the Orphan Bank Shiraz, is produced. The old vineyards can be visited – they are on the daily tours of the winery.

Peter Lehmann Wines is next door, along the banks of the North Para River. The cellar door includes beautiful and spacious picnic grounds. The history of the winery is reported elsewhere in this book. Peter Lehmann sources its grapes from 165 growers across the Barossa, and has access to 220 Shiraz vineyards, probably more than any other winery in the Barossa. It also owns a small number of vineyards, among them the Stonewell vineyard. The famous Stonewell Shiraz, however, is sourced from a number of mature and low-yielding vineyards across the valley, mainly from the north in wetter years and from around Tanunda in drier years. It is a richly concentrated wine and matured in 100 per cent new oak, 90 per cent of which is French. Peter Lehmann produces a number of Shirazes, most of which are blends. In addition, there are individual vineyard wines from Light Pass and Koonunga/Ebenezer.

The ***Chateau Dorrien*** cellar and cellar door is a landmark Barossa building. The Italian Martin family bought the site from Seppelt in 1984. The focus today is on fortified wines, but the family also produces the Black Label Shiraz, from Hoffnungsthal fruit in the Lyndoch sub-region.

The Veritas winery, now ***Rolf Binder***, is in Stonewell. Rolf Binder has been making wine for 30 years and he sells wines under the Rolf Binder, Magpie and J.J. Hahn brands. He produces quite a number of different Shirazes, including the highly acclaimed single-vineyard Hanisch and the Heysen Shiraz. These come from vines grown close to the winery, on predominantly deep sand.

Rolf Binder's mantra is to use minimal intervention in the winemaking process, but to do everything right to guide it properly on its way. This means having an instinct as well as a winemaking philosophy. There are spikes in flavour intensity at different times in different vineyards. One needs to know how the grapes mature, and to pick at the right time: 'It is key to have fruit arriving at the winery in the condition you want,' he says. An intelligent use of oak is also important. For red wines from sandy and sandy loam soils, Binder likes to use heading down boards (submerged caps) in the fermentation process. The skins of fruit from heavier soils tend to be pumped over more to increase tannin and colour extraction. His best wines are handled according to the phases of the moon, in line with the biodynamic principles outlined by the Austrian philosopher Rudolf Steiner.

The objective for the Hanisch Shiraz is to be an 'iron fist in a velvet glove'. The wine has to be big on structure, texture and weight, but have an elegant long-lasting mid palate with a rolling back palate. The Heysen is more masculine, with a more angular structure. There is more obvious French oak, but it is not dominant.

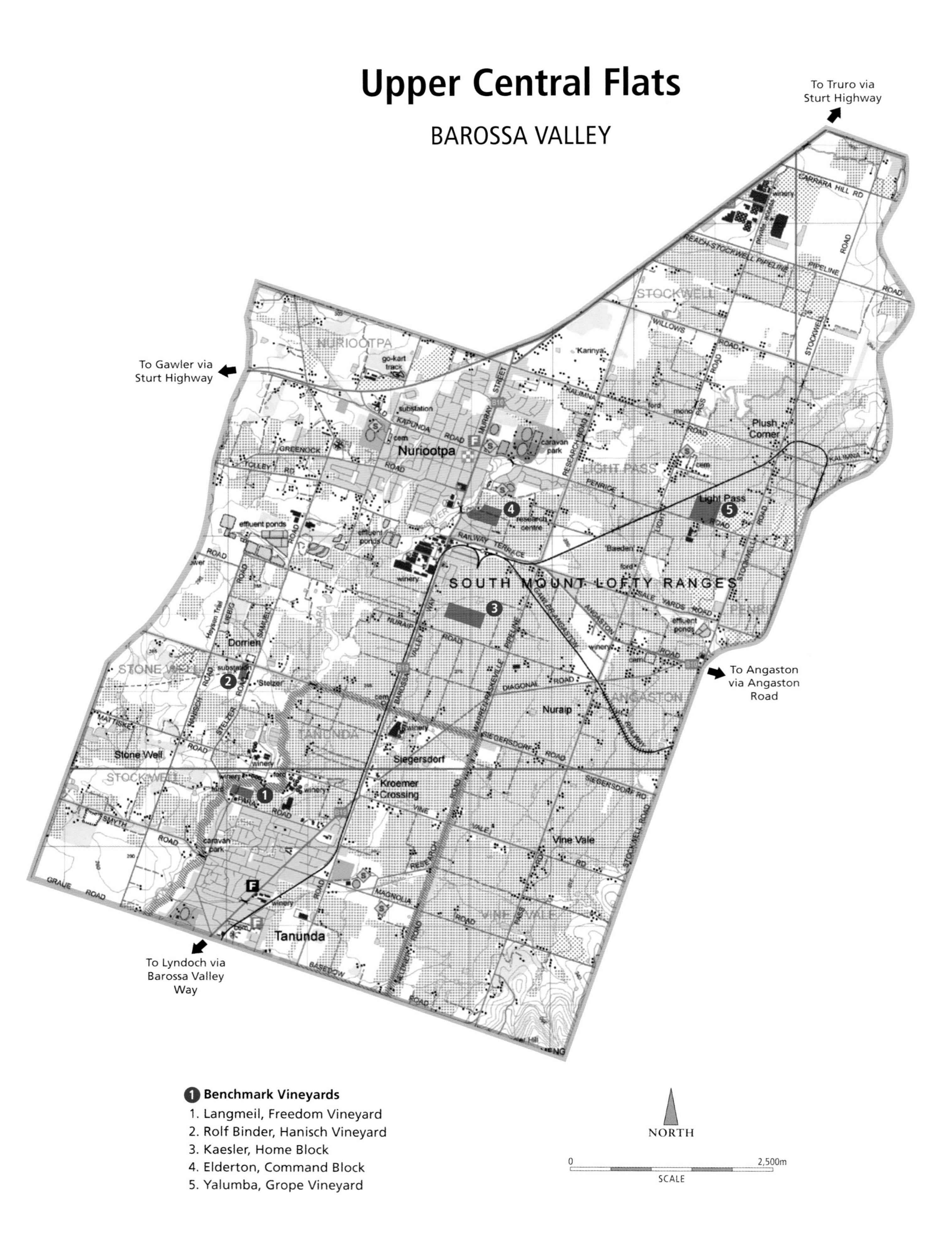

Upper Central Flats
BAROSSA VALLEY
To Truro via Sturt Highway
To Gawler via Sturt Highway
To Angaston via Angaston Road
To Lyndoch via Barossa Valley Way
Nuriootpa
Tanunda
Dorrien
Stone Well
Siegersdorf
Kroemer Crossing
Vine Vale
Nuraip
Light Pass
Plush Corner
SOUTH MOUNT LOFTY RANGES
1 Benchmark Vineyards
1. Langmeil, Freedom Vineyard
2. Rolf Binder, Hanisch Vineyard
3. Kaesler, Home Block
4. Elderton, Command Block
5. Yalumba, Grope Vineyard
NORTH
0
2,500m
SCALE

Next door is ***Whistler***. It is owned by the Pfeiffer (German for whistler) family and has a very friendly cellar door. The Shiraz and the Reserve Shiraz are sourced from the Heysen Estate Vineyard, which was planted with Kalimna 3C clones in 1994. They are matured in French and American oak for 20 months before bottling.

Just south of Nuriootpa is the ***Kaesler*** cellar door. This Shiraz specialist sources fruit for its single-vineyard Old Bastard Shiraz from the Home Block, which was established in 1893. The soil is sand over clay. Vines planted in 1961 are the source for other Shirazes. Kaesler also has a vineyard at Marananga/ Seppeltsfield.

The imposing ***Penfolds*** Barossa winemaking facility and cellar door is a little further down and across the road. There is perhaps no other winery in the world with such a range of red wines as Penfolds. Penfolds blends most of its wines across regions, although very significant proportions of the blends come from the Barossa Valley.

The two most important Grange vineyards are in the Northern Barossa, and will be addressed there. Until 2011, the only 100 per cent Barossa-sourced Shiraz has been RWT. RWT stands for 'red winemaking trial'. That name has never changed, although the wine is now well established: the 2001 was served at the Nobel Prize ceremony in Stockholm in 2005. The fruit comes from Penfolds' own vineyards and contract growers in Marananga/Seppeltsfield and the Northern Barossa, plus a small amount from Lyndoch. Penfolds has now introduced its first sub-regional Shiraz, the Bin 150 Marananga Shiraz.

Penfolds was founded in 1844 by Dr Christoph Penfold at Magill. It grew into a significant wine company soon after, producing fortified wines as well as a range of table wines. The Nuriootpa facility was bought in 1911, and this was the start of the Penfolds focus on the Barossa. In the early part of the 20th century, the emphasis was on fortified wines, but this began to change when Penfolds bought the Kalimna vineyard, then the largest in South Australia, in 1945. The history of Grange, which started in 1951, is covered in Chapter 2, as it changed the history of winemaking, in particular in the Barossa, forever. Under Max Schubert's leadership, a house style started to emerge in the 1960s across the range of red wines.

Penfolds' winemaking philosophy is all about maximising fruit quality from the vineyards (both company-owned and independent growers). Although many wines, from across regions and at different price points, are produced, and they show differences in style and character, there is 'Penfolds DNA' running through all of them. All wines show fruit and oak complexity and should improve for many years, some for decades. Though there is much multi-regional and multi-vineyard blending, thus creating a kind of 'all-round wine', Barossa Shiraz plays a major role in most wines. It is used because it adds richness, fruit concentration and length, thereby increasing the wines' ability to age well.

Red wines are still made in a traditional way, using static fermenters with header down boards. Oak plays a major role in the style. Grange is matured in 100 per cent new American oak, RWT in 100 per cent new French oak. Other wines tend to be matured in a mixture of old and new oak, but the majority use American oak. An important part of the Penfolds style is to mature wines as complete blends, not separate parcels. This is essential for a consistent house style year after year.

Elderton is another family-owned winery with an old, first-class vineyard on the banks of the North Para River. It was bought by the Ashmead family in the late 1970s and sold its first labelled estate wine from the 1982 harvest.

The winemaking philosophy here is to produce wine true to its variety and *terroir*, balanced and made for ageing. The use of oak has been reduced in the last few years, so that it now plays only a supporting role to the fruit. The company has a small, modern, winemaking facility. The wine is fermented in open concrete or steel fermenters. The maturation process can be up to three years, first in new, then used American and French oak. Blending of different parcels is done after maturation.

Elderton has achieved considerable critical acclaim and commercial success, in particular with its Shiraz wines. The flagship wine is the Command Shiraz. It uses grapes from the vines of the Command Block, part of the Elderton Estate vineyard – they are more than 100 years old. This wine is full-bodied, opulent and quite powerful. Depending on the vintage, blackberry, blueberry or dark plum flavours dominate, supported by confectionery, chocolate and vanilla flavours.

Most of the grapes for the second label Barossa Shiraz come from the Elderton Estate vineyard as well, but the vines are younger. The soil of this vineyard is alluvial sand over red and brown clay over limestone.

Rusden Wines sits in the middle of Vine Vale. Christine and Dennis Canute bought the 12-hectare vineyard, then in a run-down state, in 1979. Over time, the vineyard has been brought back to health. The Canutes were at first growers only, but they have progressed to winemaking: the first Rusden wines were made in the late 1990s. Today, the shift from grower to grower/winemaker is well established, and Christian Canute, Dennis's son, who has finished his apprenticeship at Rockford, is now producing a wide range of wines.

The philosophy of the winery starts with taking care of the land. Traditional methods are adhered to in the vineyard as well as the winery. The wines are fruit driven, because the owners believe in the vineyard and want the unique flavours the site provides to come through.

The Black Guts Shiraz is open fermented and spends only four to six days on skins. The short fermentation time ensures that the primary fruit characteristics of Shiraz are preserved. It is gently basket pressed, which results in soft tannins. It completes fermentation in French oak barriques, 20 per cent new, and spends six to nine months on its lees, completing fermentation. It then spends 30 months in barrel. The gentle treatment and long maturation deliver a softened, well-integrated wine with a silky feel on the palate. This wine tends to be lighter in colour and more focused on texture than fruit expression.

The vineyard has deep white sandy soils over red clay: this is typical for Vine Vale. It gives the Shiraz lifted characters of violet and white pepper. The gully winds are responsible for the late ripening and relatively high acidity of the wines. The cooler climate and deep white alluvial sand result in a unique *terroir* here.

Yelland & Papps is perhaps the odd one out in this list of wineries, since most of its wines have been associated with the Greenock sub-region, but in 2010 Susan and Michael Papps purchased a property on the outskirts of Nuriootpa, where they have a 1-hectare vineyard, and where they have established their winery and tasting room. The other unusual aspect of this venture, for the Barossa, is that neither has a history of wine in their family. Both, however, learnt from many winemakers in the Barossa before they started to make their own wine in 2005.

Like many others, their stated winemaking philosophy is to showcase the essence of the vineyard, intervene as little as possible and allow the fruit to speak for itself. What stands out is their passion for making good quality wine – and their ambition to grow.

The two original wine ranges are Delight and Devote; the more upmarket Divine and Decadent ranges have since been added. The Devote Greenock Shiraz is their main Shiraz at present. The fruit comes from a David Materne vineyard in Greenock, planted in the late 1980s. The wine shows typical characteristics for the area: rich dark berry flavour with a firm tannin structure and a fleshy long finish. The Divine Shiraz is from older vineyards, has longer skin contact (for added extraction) and longer oak maturation.

John Duval Wines are also listed here, as they are sold from the new cellar door of the 'Artisans of the Barossa', which previously belonged to Murdock Wines. John Duval's contribution is covered in Chapter 5, but I will deal with his wines here. The business started in 2003/04 with the release of the Plexus GSM and the Entity Shiraz, which were followed by the first release of the Eligo Reserve Shiraz in 2005.

Duval's philosophy is to let the vineyards express themselves in an elegant fashion. He picks the grapes before they are too ripe and tries to avoid high alcohol levels. He looks for good acidity to maintain vibrant fruit expression and to achieve a good marriage with food. Fine-grained French oak is used to support the fruit. Oak maturation is there to soften the tannin structure, but wine that is left too long in barrel can flatten, and lose fruit intensity, so bottling at the right time is also important. Duval uses a variety of fermenters in order to have a choice of fermentation techniques. Otherwise, the process is quite standard.

The John Duval Shirazes are full-bodied, savoury rather than jammy, with good concentration of black fruits and dark chocolate. The tannins are ripe and fine. The Eligo has a more layered palate and a structure and fruit intensity that suit longer term cellaring. Duval has access to quite a few vineyards, and the fruit mainly comes from the Western Ridge, the Northern Barossa and Eden Valley.

The Seabrook name has been part of Australia's wine business since 1878. It has been mainly a wine merchant business, but finally fell dormant. A number of family members were distinguished wine judges – in fact five generations have judged at the Royal Melbourne Wine Show. Hamish Seabrook is one of them. He was educated at Roseworthy and was winemaker at Bests in Victoria before he launched ***Seabrook Wines*** in the Barossa in 2005.

The Seabrook Barossa Shiraz, called the Merchant, is perhaps not typical for the region. The forward fruit is restricted and the focus is on structure and ageability. Half the fruit for the Shiraz comes from Paul Clancy in Krondorf. This vineyard is on red dirt and produces wines that are fruit lifted, luscious and rich. The flavours of wine made from this vineyard are blackberry and tapenade. The vineyard crops 5 tonnes per hectare and is roughly 30 years old. It is at the base of the Barossa Ranges. The other 50 per cent comes from the family vineyards at the bottom of Mengler Hill in Vine Vale. The soils are red–brown clay over white clay. Wines from this vineyard have great structure and poise; they have dense richness and are savoury through and through.

The ***Sons of Eden*** winery is at Light Pass. It takes its name from its two partners, Simon Cowham and Corey Ryan, both of whom have learnt and refined their skills in the vineyards and cellars of Eden Valley, including Henschke, Heggies and Pewsey Vale. They started this winery in 2000, after having gathered experience at those other leading wineries.

The two key Shirazes are the Romulus and the Remus. In the same way that Remus was the refined, focused and elegant brother of the legendary founders of ancient Rome, the Remus wine comes from low-yielding old vines at Eden Valley and has a soft and refined palate based on dark cherry fruit, with lifted aromatics; it also has great fruit concentration and a rigid backbone. The Romulus is made from old vine fruit at Light Pass and Moppa. Light Pass, the 'home' vineyard – red loam over red clay over limestone soil – contributes fruitcake flavours, and Moppa contributes great colour, tannin structure, and licorice flavours. This is the more powerful of the two wines, backed up by vanilla chocolate from the American oak.

The Willows Vineyard is also at Light Pass. The property has been owned by the Scholz family since 1845, but the vineyard was first planted in 1936. Own-label winemaking started in 1989.

All wines are sourced from this single vineyard. The aim is to achieve intense fruit flavours and ripe tannins in wines that are not big and alcoholic. The fruit goes through extended maturation in at least 50 per cent used French oak hogsheads, as the objective is to produce a generous and matured wine.

The Bonesetter Shiraz is its premium single-vineyard Shiraz. It is traditionally made in small batches, fermented in open fermenters, hand-plunged and basket-pressed.

The grapes come from vines planted in 1969. The vineyard is next to the North Para River, and the soil varies from alluvial sands to red–brown loam over clay.

Across the Road is ***Gibson Barossavale/Loose End***. Rob Gibson was a senior viticulturist with Penfolds, where he introduced a grading system to match vineyard characteristics to the requirements for Grange. He started his own winery, Gibson Barossavale, in 1997. With a focus on viticulture, he searched out old vines. He ended up with vineyards in Stockwell/Light Pass and Eden Valley. The Shirazes are mostly sourced from these vineyards, but are not strictly estate grown nor single vineyard.

Gibson's winemaking philosophy starts with picking only optimally flavoured grapes. He believes wine has an innate durability and he aims to make a natural product 'by substituting intelligent management for remedial inputs'. He applies selected yeasts to 80 per cent of the ferment, and uses wild yeasts for the rest.

The Australian Old Vine Collection Barossa Shiraz and the Old Vine Collection Eden Valley Shiraz are the flagship wines. These are produced in low volumes from very old vineyards (150 years in the case of the Barossa). They have rich fruit and tobacco flavours with savoury characteristics and dry tannins.

Wolfgang Blass established ***Wolf Blass*** in 1966. The winery has its large visitor centre off the Sturt Highway. This company is probably the antithesis of what this book is about. It is foremost a brand and marketing company. The source of the fruit is not acknowledged. Wines are blended across varieties, vineyards and regions to create a particular brand and style. The company has many vineyard holdings in the Barossa, in particular in the Central Valley, but does not publish which wines the fruit goes into.

However, things are slightly different when it comes to the ultra premium Platinum Shiraz, which aims to combine power and elegance. The fruit for this

Mengler Hill

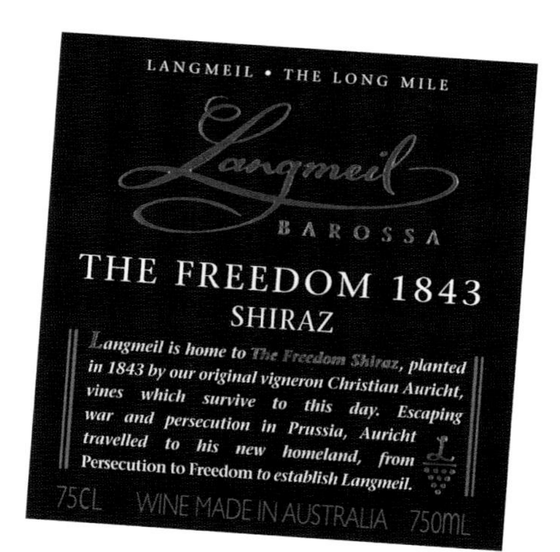

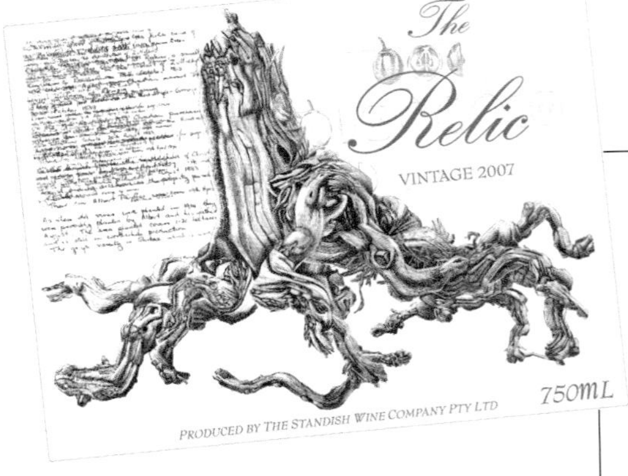

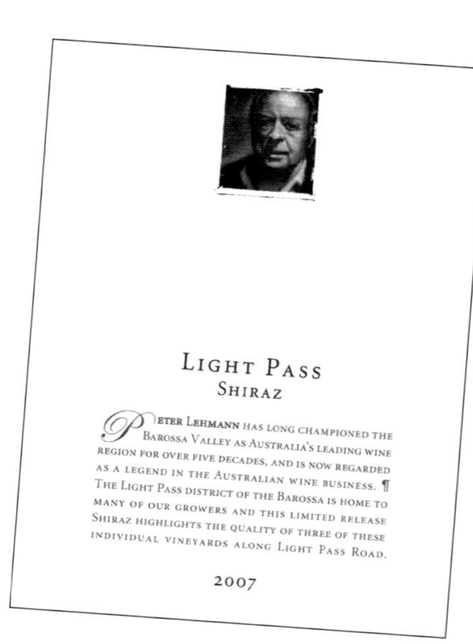

wine used to come from vineyards near Moculta in Eden Valley and the Central Valley. It now comes predominantly from the Medlands Orchard Block vineyard in the Upper Central Flats, which has 12 hectares of Shiraz, most of it planted in 1997 and 1998. The company thinks that the fruit from this vineyard, which is vibrant and intense, with supple texture and structural balance, best achieves the objective.

Fermentation, after crushing and de-stemming, takes place in small open-top fermenters which use a combination of plunging and pump-over techniques to maximise extraction. Individual parcels are fermented separately on skins until dry, generally for up to 14 days, taking advantage of time to extract colour and flavour. After gentle pressing, the wines are transferred to 100 per cent French oak barriques (70–90 per cent new) for approximately 20 months' maturation. Final blending takes place just before bottling.

The Gold Label Shiraz is the company's second Barossa Shiraz; it is made for everyday drinking.

Tasting profile

The above sections show that there are more single-vineyard wines made in the Upper Central Flats than in the Lower Central Flats. Therefore I can evaluate the tasting profile of this sub-region by comparing finished wines and barrel samples.

Like those of the Lower Central Flats, the Shiraz wines of this area do not feature very high extraction and strong tannins. The old vineyards on the alluvial soils of the North Para River produce wine with intense black fruit characteristics, but also elegance. The grapes do not need to get too ripe to provide these flavours. And the ripening occurs as much through moisture as through sunshine and heat. There are also broad flavours, including citrus and meatiness. These wines can be a little perfumed.

Plum flavours dominate in Vine Vale. The wines are as elegant as the Para River wines and lighter than those from the Western Ridge or the Northern Barossa. Vigour can be an issue in wetter years and can reduce fruit intensity in the wine.

As we move further north, the wines get richer, featuring strong plum and blackcurrant fruit, and they start to develop some chocolate flavours.

Wines from Light Pass are generous as well, but the fruit spectrum is more about red berries; the wines also show some spiciness. This is probably due to the proximity of this area to the Eastern Slopes and to the gully winds which come down the valley in the afternoon – these have a cooling effect.

There seem to be many differences, but it still makes sense to group wines from these different areas into one sub-region: the emphasis of the excellent wines from this area is on fruit, elegance and balance. Tannin structures remain in the background.

And then there are the wines from Stonewell, northwest of Tanunda. This area seems like a transition to the Western Ridge. The wines are darker than in the other areas of the valley floor, quite intense, and have fine silky tannins in the background. Rolf Binder's Hanisch is a good reflection of what goes on here. There are four different soil types that run through his vineyard: sandy soils, sandy loam, sand over ironstone and black ironstone. This may explain both the complex fruit and the very long tannins. It is a bit like this for the whole area, although it is really difficult to establish exactly the factors causing this change from the valley floor: the Heysen, for instance, which comes from fruit only 200 m away, does not have the same intensity and oomph.

Vineyards

Benchmark vineyards in the Upper Central Flats are the Freedom Vineyard, Kaesler's Home Block, Elderton's Command Block, Binder's Hanisch Block and Yalumba's Grope Vineyard at Light Pass. The Freedom Vineyard wines express the black fruit characteristics and elegance which can develop in very old wines along the North Para River. The Home Block and Command Block show increasing intensity and palate weight. The Hanisch Block reflects the complexity of its soil and the transition to Marananga/Seppeltsfield. The Grope vineyard includes 90-year-old vines, grown on fine sandy loam over red–brown earth. It has consistently been a major contributor to the Yalumba Octavius and The Reserve, and delivers dark fruit concentration and fine long tannins.

Eastern Slopes

The Eastern Slopes as a special sub-region was first written about by David Farmer, a geologist who has been studying sub-regionality in the Barossa Valley.

Terroir

A fault line on the eastern side of the Barossa Valley emerged a few million years ago and formed what is known as the Eastern Range. Some parts of this, in particular north of Angaston and south of Rowland Flat, are covered by thick sediments of pebble and bright red soils, which have been washed off the Eastern Range. In the higher parts of the Ranges, the soil has a lighter texture and there are bare rocks and steep slopes. Often, the ground is not suitable for vineyards. Eden Valley starts east of the ridge.

I have defined the western border of the Eastern Slopes as basically the Stockwell Road in the north and a 340 m elevation line across the south. The vineyard opportunities are mainly in the north. The elevation can be up to 420 m. Higher elevation and special soil conditions are not the only things that make this area special. The other factor is the cool gully breezes that come down the slopes late in the afternoon during summer. This combination leads to a more gradual ripening process than on the valley floor. Vineyards here are often planted to Riesling, as these conditions are generally unsuitable for Shiraz. The main opportunities for Shiraz are west of Angaston, leading to Eden Valley.

Major wineries

There is only one winery with a major cellar door in this sub-region: ***Saltram***, near Angaston. It goes back to 1844, when William Salter purchased his land parcel. The Mamre Brook house, in which he lived, still stands today. Today, Saltram is owned by Treasury Wine Estates, which has recently changed the branding of premium Shiraz. The first wine, called No. 1 Shiraz, was originally made in 1862. The No. 1 Shiraz is now being made again, and the fruit comes from the Saltram Vineyard, which surrounds the cellar door and has vines dating back to the 1950s, as well as from other vineyards, mainly in the Eastern Slopes. The second wine is the Journal Shiraz, which includes a significant Eden Valley component.

Tasting profile

It is too early to describe exactly and consistently the flavours of Shiraz from the Eastern Slopes, as no wines are made from individual vineyards here. A couple of samples I have tried suggest that the cooler temperatures lead to wines which have less deep colour and are not as big and juicy as those from the valley floor. They have perhaps more red fruit character than the plush plum flavours from nearby, and include some earthy characters and higher levels of natural acidity.

Eastern Slopes

BAROSSA VALLEY

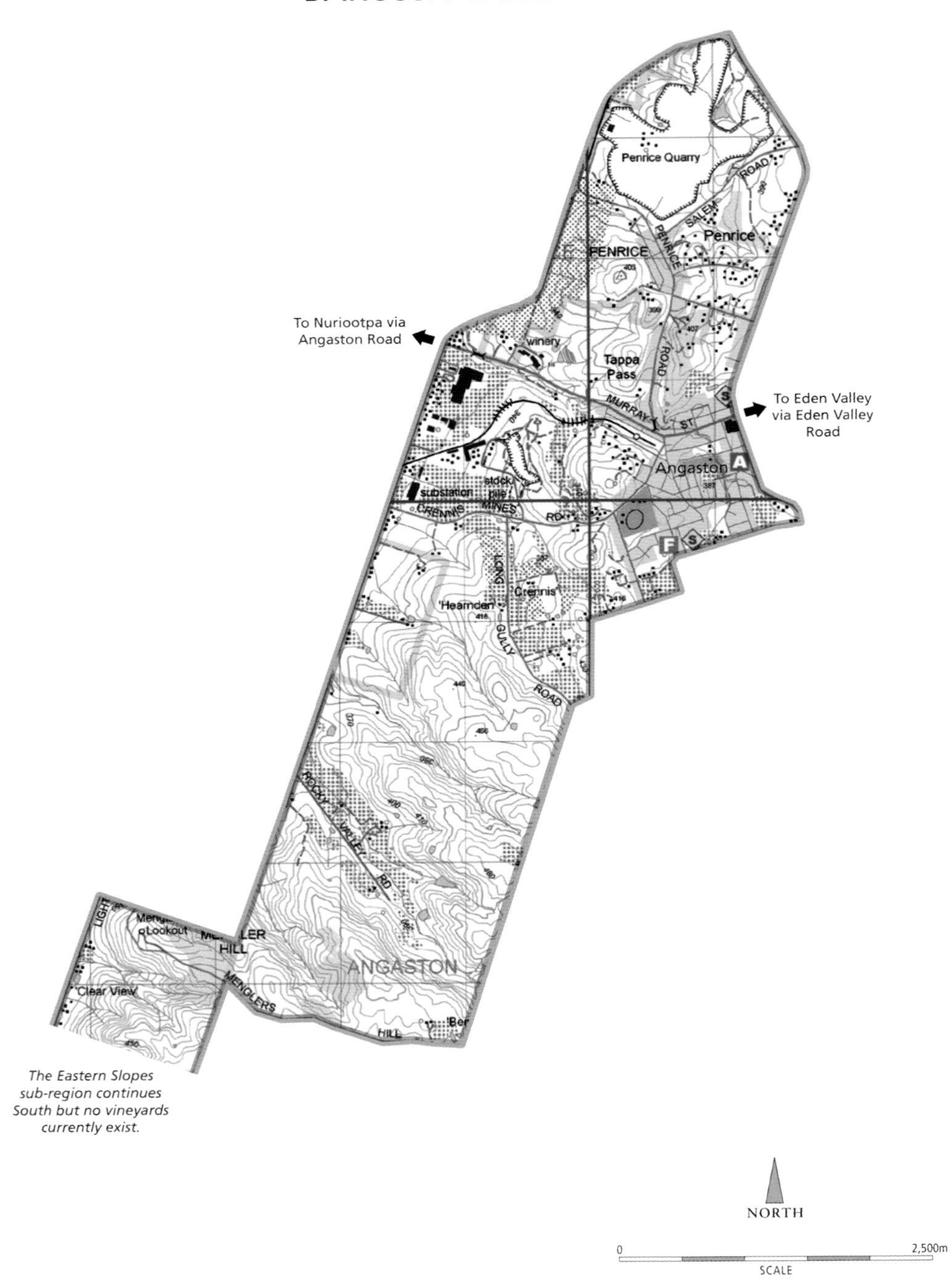

The Leaders of the Barossa, Part 1

Introduction

Terroir is not great by itself; it needs to be found, nurtured and brought to the attention of others. This requires vision, leadership, hard work and persistence. Throughout this book, I have added short portraits of those who have been mentioned as leaders most often in my interviews with winemakers and viticulturists over the last 30 years.

I have sorted these leaders into three groups. The first group are those who believed in the Barossa during the dire times – the early 1980s. They laid the foundation for the great Barossa Shiraz wines we enjoy today. The second group took the Barossa to new heights a decade later, and the final group represents the next generation, the young guns. What do they all have in common? Conviction: conviction that they can do something that has not been done before in the Barossa. Their innovations will create new legacies.

The selection can of course be debated. Some people have narrowly missed out – Rob Gibson, with his viticultural work and the introduction of the Penfolds payment system, and Damien Tscharke, with his focus on alternative varieties, spring to mind as examples. There is no doubt that some other young guns, for example from the 'Artisans of the Barossa' group, will also make their mark in the future.

The Peter Lehmann team

'In 1977 I got this missile that arrived by post.' The letter instructed Peter Lehmann, who was working for Saltram at the time, to tell growers that Saltram – one of the biggest customers for Barossa grapes – would not be buying any 1978 fruit. This was a bombshell, because Lehmann had previously been told to solicit grapes: 'I'd given the growers my word. I was damned if I was going to break that.'[4]

Peter Lehmann was a fifth generation Barossan who valued trust, loyalty and friendship, and he had unshakeable faith in the Barossa. With money from family and friends he stood by his growers, bought the fruit, and in 1979 built the Masterson Barossa Vignerons winery. His determination saved many old vines; he also encouraged others to form family wine companies, based on old vine Shiraz. Peter Lehmann became the doyen of the Barossa. In 1982 he sold his wine under his own name, Peter Lehmann Wines, for the first time. He then steered his company through a number of ownership changes, and the company has continued to flourish.

4 Quoted in Felicity Carter, 'Peter Lehmann: Vintage Barossa', *Meininger's Wine Business International*, 10 April 2008.

Margaret and Peter Lehmann; Andrew Wigan (left); Doug Lehmann (right)
[Photos: Peter Lehmann Wines]

Lehmann understood what the Barossa does best and believed in the Barossa as a premium area for Shiraz. He took his Barossa Shiraz overseas and won many awards, culminating in the Lifetime Achievement Award at the prestigious 2009 International Wine Challenge Awards in London.

Andrew Wigan, Peter Lehmann's winemaker for 30 years and originally a grower/supplier himself, has known for some time that every year there are some parcels of grapes that are more intense, more muscular, deeper in colour. He calls them 'little black jewels'. In 1987, he and Lehmann created Stonewell with these grapes. Each year, a selection of these special grapes goes into this wine. Stonewell is a wine built for ageing, and it has given further credibility to Barossa Shiraz.

Peter's son Doug took over as Managing Director in 1990. He has continued to increase the company's portfolio of wines – not only reds, but also Riesling and Semillon – to the point where the grapes from 180 growers are now converted into 750,000 cases of wine. Grapes come from many parts of the Barossa, with perhaps a focus on Light Pass. Under his leadership, the company is an efficient producer, and delivers good value for money at every price point.

The true legacy of this leadership team is its relationship with its growers, the creation of 'brand Barossa' and the team's great personal knowledge of many vineyards across the Barossa. Peter, who died in June 2013 aged 82, will be sorely missed.

Robert O'Callaghan

'One of the main reasons for setting up Rockford was to attain the independence to make wines that are not necessarily fashionable at volume retail level. Two wines in this category are about to be released under a special label … Then there is the 1984 Basket Press Shiraz … and to quote the back label of this release … "Uniquely Australian, it's loaded with rich, ripe fruit flavour and very fine tannin".'[5]

Robert O'Callaghan
[Photo: Dragan Radocaj]

The order form from 1986/87 is interesting, not just because it shows how prices have increased, but also, in particular, because the price of the Basket Press Shiraz is not so much more than that of the other wines.

This wine did not remain unfashionable for long. Within a couple of years it became Rockford's flagship wine. In the Krondorf area, a small number of winemakers wanted to save the old Shiraz vineyards; Robert O'Callaghan was one of that group's leaders. In 1980 he introduced Old Block Shiraz at St Hallett, and in 1984 he created and launched Basket Press Shiraz at Rockford.

In hindsight, this period was an ideal one in which to launch a premium Shiraz: nobody wanted the grapes from these old vineyards, so they had fallen dramatically in price. O'Callaghan, over time, negotiated exclusive contracts with 30 of these vineyards by paying a premium price for these grapes. This

5 Robert O'Callaghan, *Rockford Winery Newsletter*, 1986.

ORDER FORM Product	per Bottle	per Ctn. (Less 10%)	Quantity Ordered	Value $ c
1986 Rhine Riesling Eden Valley	5.60	60.50		
1985 Sauvignon Blanc – Adelaide Plains	6.90	74.50		
1986 White Frontignac – Spaetlese	5 60	60.50		
1985 Rhine Riesling – Botrytis Cinerea 375ml	5.90	63.70		
1986 Alicante Bouchet – Adelaide Plains	5.90	63.70		
1981 Shiraz/Cabernet – Barossa Valley/McLaren Vale	5.60	60.50		
1982 Cabernet Sauvignon – McLaren Vale	6.90	74.50		
Muscat of Alexandria 750ml	7 70	87.80		
375ml	5.50	62.80		
Tanunda Tawny Port – 12 years old	10.50	113 50		
1975 Shiraz Vintage Port	7 70	87.80		
1984 Basket Press Shiraz	8.25	89 10		
1986 1886 Vine Vale Rhine Riesling	8.25	89.10		
All wines listed include License Fee and 20% Sales Tax			Total	

approach was based on his understanding of the quality of these vineyards; they were being ignored by most other winemakers at the time.

Robert's second major achievement was managing, as the founding Chairman of the Barossa Residents' Association, to change the state government's view of the region. The threatened subdivision of land in the area did not occur, and apart from the villages, all other areas were rezoned rural. They are still zoned rural.

The Rockford winery stands out because it has a strong philosophical framework, based originally on Lutheran principles about preserving the cultural heritage of one's community. The winery has remained a traditional winery in more than one sense: it uses traditional wooden equipment, and it focuses on direct contact with its customers. It has grown slowly, and has maintained strong links to growers and customers. Its grower dinners and customer lunches have become legendary.

Robert O'Callaghan is one of the most passionate winemakers anyone could ever meet, with strong convictions and great leadership skills; he is also a very successful winemaker. The Basket Press Shiraz style – full-bodied, ripe wine with a long-lasting finish – became the model for the big Barossa wines which became fashionable in the 1990s. Robert invested in training, and some of today's best-known winemakers in the valley, including Chris Ringland, Dave Powell, Stuart Blackwell and Andrew Seppelt, come from the 'Rockford Academy' (his term).

Charles Melton

Charles Melton is another winemaker who set out on his own in 1984, believing in the quality of the old vines of Rhône varieties, in particular Shiraz, Grenache and Mourvèdre. His first small vineyard on Krondorf Road grew Grenache bush vines, and this has been his flagship grape variety ever since.

Charlie – so named by Peter Lehmann: his original name is Graeme – came from Sydney to the Barossa in the 1970s, and started as a cellar hand at the small Krondorf winery. He later moved with Peter Lehmann to his Masterson winery, where he learnt to make full-bodied red table wine. He then started out on his own.

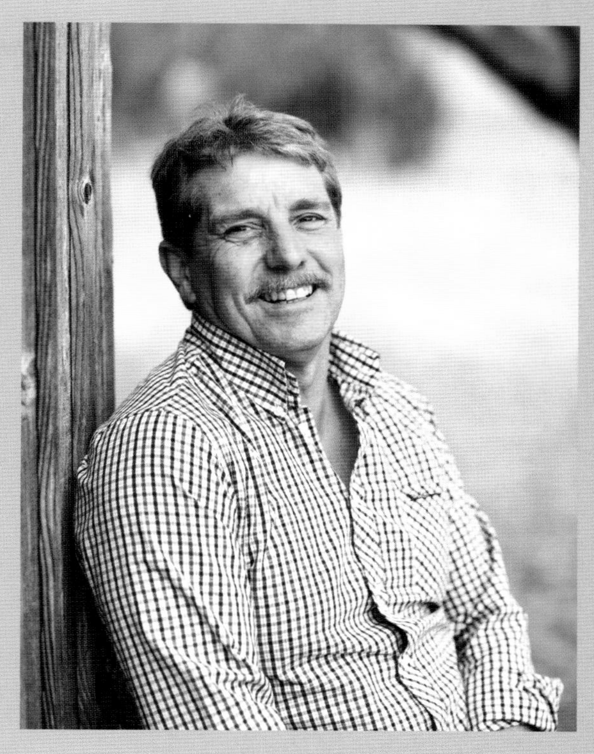

Charles Melton
[Photo: Randy Larcombe]

He is particularly famous for his Grenache/Shiraz/Mourvèdre blend, Nine Popes, which is probably still the most highly regarded blend of this kind in Australia. The story of how the wine got its name is now legendary. As he was sitting at a bar with a friend trying to find a name, they were looking at the Châteauneuf du Pape area as the model. When translated to English, this means 'new castle of the Pope', and it refers to the alternative papacy in the 13th century. However, they thought 'neuf' meant 'nine', and so Nine Popes was born. Melton was obviously not too familiar with French, or with papal history.

Like many of his colleagues, Melton focused on older vineyards, and Rhône-style wines. Besides Nine Popes, his Rosé – Rose of Virginia (another name twister: this relates to a flower, not the wine variety) – became a model that many winemakers have imitated.

His contribution to the growing and making of Rhône-style wines in the Barossa is great, and he is recognised for it by his peers. He is also valued for his appreciation of the culture and history of the area – food, music, art and religion – and the commitment he has made in these fields.

Bob McLean

Bob McLean has a reputation for never missing lunch. Yet it was during the Wednesday nights – not days – spent at Maggie Beer's farm with Peter Lehmann, Robert O'Callaghan and Charlie Melton in the 1980s that this group of smaller winemakers gathered the courage to literally 'bank' on the Barossa and old vine Shiraz. In 1987, McLean had the chance to buy into St Hallett. It took him 10 years to pay off his debt, but he did it – by growing St Hallett into a sizeable company during a very difficult time in the industry.

Bob McLean

Before this, McLean had worked in public relations at Orlando – from 1972 to 1985. In that position he was a major force behind developing the Jacob's Creek brand. He further developed his marketing skills at Petaluma, from 1985 to 1987.

McLean is particularly appreciated by his colleagues for those marketing skills, which he successfully applied again when he was rebuilding St Hallett as part-owner and Managing Director.

Today, St Hallett is owned by Lion-Nathan, a large corporation, but the philosophy remains: the focus is still on three-tiered Barossa-focused Shiraz based on mature vines. The Old Block, which is always made from vines that are at least 60 years old, and comes from Lyndoch and Eden Valley vineyards, is at the top; it is a wine that you 'can drink a whole bottle of without your tongue ripping out' (Bob McLean at a 2004 wine tasting). It is complemented by the Blackwell, which McLean introduced, and which uses grapes from Greenock, Marananga/Seppeltsfield and Koonunga/Ebenezer, and the entry level and fruitier Faith Shiraz.

Today McLean runs his small family winery, McLean's Farm, in Eden Valley.

The Southern Valley

5

The Southern Valley is the second major section within the Barossa Valley. It is different from the Central Valley in a number of ways. First of all, it has a different soil profile. The area forms an alluvial plain, but the sediment consists of finer sands than north of Rowland Flat. It covers schists, siltstones, calcilicates and quartzites. Also, there is some Barossa ironstone in this area, mostly on the slopes of the western side. In the Williamstown region, some areas are red–brown earths, while others are quite sandy.

The rainfall in the Barossa increases from north to south, with Lyndoch having 5–10 per cent more rain than Tanunda, and Williamstown an additional 10–15 per cent. However, the number of rain days in Williamstown is not higher, which indicates that the falls there are heavier.

The Lyndoch Valley is at a lower elevation than the Central Valley: most is between 170 m and 220 m, with Gods Hill going up to 250 m. The vineyards near Williamstown are at an elevation of 250–300 m.

Lyndoch

The Lyndoch Valley sub-region consists of a number of different areas, complexities and mesoclimates. It starts south of the Para River and is almost a square area, including from Hermann Thumm Drive to Chateau Yaldara, the main valley towards Williamstown, Gods Hill, and Hoffnungsthal. It stretches to the border of the Barossa Valley in the west. This area, up to Rosedale, may form its own sub-region in the future. At this point, the vines planted in the west are quite young, and much of the area is owned by groups of people or companies – the vines are part of their investment schemes. The destination of the grapes remains mostly unknown. I am extending the Lyndoch area south to include Cockatoo Valley and the vineyards along Mamba Road, as the *terroir* there is closely related to that of the other parts of Lyndoch.

The centre of this sub-region is the township of Lyndoch, first explored by settlers in the late 1830s as they moved to the Barossa Valley. Today it has a population of approximately 1500. It is mainly a service centre for the wine industry. There are several wineries, plus accommodation and restaurants, near Lyndoch.

Terroir

The Lyndoch Valley has the lowest elevation in the Barossa Valley. As a result, it is warmer than other areas. As mentioned above, there is usually higher rainfall

Lyndoch

BAROSSA VALLEY

❶ Benchmark Vineyards

1. Grant Burge, Filsell Vineyard
2. Dutschke, St Jakobi Vineyard
3. Rick Burge, Draycott Vineyard
4. Torbreck, Hillside Vineyard
5. Yalumba, Fromm Vineyard

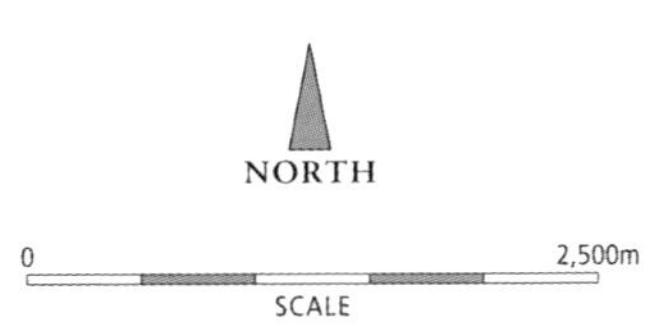

here than in the Central Valley. The soils are quite complex: the area northeast of Lyndoch, towards Altona, has very sandy soils. North of Lyndoch, towards Gomersal, there are red–brown earths and grey loam of a heavy texture. Similar soils and alluvial soils dominate in the flatlands towards Williamstown. Barossa ironstone is found throughout the Lyndoch area. There is also a lot of sand on Gods Hill.

As a result of this complexity, soil conditions vary from vineyard to vineyard. However, a number of generalisations about Lyndoch are possible. The main one is that vines ripen early in this area, mainly because of the lower altitude and its resulting increased warmth. There can be quite a lot of vigour, particularly in wetter years, so canopy management is important.

Major wineries

There are a number of major wineries with cellar doors in the Lyndoch sub-region. At the northeastern end is **Kellermeister**. Its focus is on working closely with growers to understand the specifics of each site. Viticultural decisions are seen as just as important as winemaking decisions. Old vines are always hand picked, and picked on flavour, not chemical analysis. Kellermeister has different fermenter types: they are different sizes, and there are some with and some without header boards. This allows extractive fermentation and cap management techniques to be tailored to each individual wine batch.

There are two premium Shiraz wines. The first is the Kellermeister Black Sash Shiraz, a single-vineyard wine. In most years it is made from 130-year-old vines at the front of the winery and matured in French oak. The second is the Wild Witch Shiraz, a blend of Northern Barossa and Eden Valley fruit. It is characterised by blue and black fruits, with an underlying earthy complexity and a silky palate.

Schild began as a grower, and started to produce wine under its own label in 1998. It has grown strongly since then. Schild has more than 160 hectares of vineyard, but most of the grapes are still sold to other wineries. All its vineyards, with the exception of Moorooroo in the Lower Central Flats, are in the Lyndoch area. The Shirazes are, with the exception of the 2008, estate grown, and include the Barossa Shiraz, the Ben Schild Reserve Shiraz and the Moorooroo Limited Release Shiraz. The Schild cellar door is in the centre of Lyndoch.

The Moorooroo Limited Release Shiraz is produced from what is left of the vines that were planted by Johann Gramp in 1847, which makes them some of the oldest Shiraz vines in the world. 'Moorooroo' is an Aboriginal word which means 'meeting of two waters'. It refers to the North Para River and Jacobs Creek – the vineyard is just north of Jacobs Creek. The wine is matured in American oak hogsheads, following extended time in contact with the skins during processing to maximise the colour and flavour extraction. The wine is bottled with some filtering. It is a good example of the intensity and complexity that low-yielding old vines can produce.

The higher volume Barossa Shiraz comes predominantly from Schild's Lyndoch holdings. It usually has ripe cherry characteristics with vanilla from new and old American oak.

Another winery that processes only estate-grown vines is **Burge Family Winemakers**. This family winery has been making wine since 1928. During the 1960s, the business incorporated as Wilsford Pty Ltd and was owned 50 per cent by Rick Burge's father Noel and 50 per cent by Grant Burge's father Colin. The winery was well known for its fortified wines and continued with them into the 1980s, when the sons took over, but with increasingly less fortune. In 1986 Rick Burge took the plunge and bought out Grant to form Burge Family Winemakers. The objective then became to produce table wines, mainly from the key Rhône varieties of Shiraz, Grenache and Mourvèdre, but also some Riesling and Semillon.

Rick Burge's overall philosophy is to work with nature. He sees himself as primarily a wine grower. He uses no chemicals on the soil and pays attention to the phases of the moon – different phases are judged as particularly suitable for planting, cultivating or harvesting, according to biodynamic farming principles. His objective is not to make blockbuster wines, but he cringes at the anti-blockbuster sentiment that has become fashionable since the mid 2000s. His wines are generously – and long – flavoured; there is no need to work the skins to enhance the colour of the wine or the intensity of the fruit.

The two main Shirazes, Draycott and Olive Hill (the latter includes some Grenache and Mourvèdre), are single-vineyard wines from vineyards surrounding the cellar door, which is just north of Lyndoch. The 4-hectare Draycott vineyard has loam over red clay. Some vines from around 1960 remain, but most Shiraz was replanted after 1986. The 4-hectare Olive Hill Block, west of the winery, has complex soil, mainly alluvial, limestone, red clay and some quartz. It has some 90-year-old Grenache, but again, most vines were planted after Rick Burge took control. Both wines are aged in about one-third new French oak barrels and two-thirds used French oak barrels. Burge treats oak as a seasoning agent, and uses it to achieve longevity, not to affect flavour. Therefore oak maturation is only 12–15 months. Burge continues to bottle these wines with cork. He believes that cork is the superior closure for bringing out the ageing characteristics inherent in premium red wines.

Across the road is **Kies Family Wines**. The Kies family migrated to Lyndoch in 1857 and has been growing grapes there ever since. Most of the current vines in the estate vineyard, which is behind the cellar door, were planted in 1965.

The cellar door has been upgraded over the last 20 years and an attractive café has been added. The winery is now managed by the fifth and sixth generations of the family. Kies is best known for its Merlot, but also produces the Dedication Shiraz, from 80-year-old vines, and the Klauber Block Shiraz.

Chateau Yaldara, owned by McGuigan Wines, is at the northern end of the Lyndoch area. It is an impressive and picturesque complex that includes a café, delicatessen and conference facilities. The main sandstone-fronted baroque look-a-like building, which serves as the tasting room, was built in 1947 by German winemaker Hermann Thumm. The wines available for tasting include McGuigan wines from other regions.

The Standish Wine Company is being established in the northeastern hills of Lyndoch. The company was founded by Dan Standish, a sixth generation grapegrower, in 1999. He learnt winemaking at Torbreck, as well as at wineries in California and Spain. The first wine he produced, the Relic, which includes 4 per cent Viognier, comes from the old family vineyard in Vine Vale, planted on own roots in 1912. Since then the portfolio has grown to five single-vineyard Shirazes, all from dry-grown, low-yielding vineyards. The Standish is a 100 per cent Shiraz from the same vineyard, the Andelmonde is from a vineyard in Greenock, the Borne Bollene is from the Fechner vineyard in Eden Valley, and the latest, the Schubert Theorem, is from the Schubert vineyard at Marananga.

Dan Standish embraces the specifics of each *terroir*, vintage and clone. His aim is for the wine to show what the vine plant has experienced over the growing season, as well as the qualities of the soil and micro-climate – he wants to create wines with a true and pure sense of place. As an example, he sees the fragrancy and purity of fruit from the Vine Vale wines as a reflection of its sandy soils.

Accordingly, the winemaking technique allows for minimum intervention, starting with dry-grown plants, low yields and hand harvesting. Standish uses native yeasts and no additives. The fruit is gently basket pressed, has extended lees contact, and there is no filtration or fining.

Dutschke Wines is located on Gods Hill Road. Wayne Dutschke started to produce wine in 1990 under the Willow Bend label. In 1998 the brand was changed to Dutschke Wines. For many years Dutschke worked as a winemaker for various wineries, and borrowed some space in them to store his own barrels and prepare his blends. In 2004 the decision was made to build the Dutschke winery.

Wayne Dutschke takes fruit from three neighbouring vineyards on Gods Hill Road as well as the old 55-hectare family vineyard – one-third was planted to Shiraz in 1975 – around the St Jakobi church on the Lyndoch Valley Highway, 2 kilometres south of Lyndoch. His focus is on Shiraz and fortified wines. He uses a typical small-volume winemaking process. Wines are matured in oak for 18–20 months before being blended and bottled.

The two flagship wines are the St Jakobi and the Oscar Semmler Shiraz. They are produced from the same special 2.5-hectare section of the St Jakobi vineyard. The St Jakobi comes from the east and middle section. It is matured in French and American hogsheads, with 25 per cent of the barrels being new. The wine is full-bodied, with spicy dark cherry fruit characteristics. The Oscar Semmler comes from the western part of the vineyard, where the yields are lower and the fruit is riper, and with more concentration. This wine is matured in a combination of new and 1-year-old 100 per cent French hogsheads. This wine has a rich style, with big sweet fruit and a bigger tannin structure than the St Jakobi.

Tasting profile

Some of the wineries in this sub-region produce single-vineyard Shirazes, and some produce wines sourced only from Lyndoch. In addition, I had the opportunity to taste barrel samples from Lyndoch vineyards.

The wines tend to be quite ripe, with lush fruit flavours, but not overly big. Fruit flavours vary from raspberry to (mostly) dark cherry and plum. The wines are aromatic, perfumed, soft and often quite elegant. These fresh and aromatic flavours are probably due to the higher levels of humidity here than in the Central Valley. Dark chocolate flavours are often present. Lyndoch wines show more spice than those from other sub-regions of the Barossa, and they are silky smooth. Tannins are fine, but take a back seat. One might say that the flavour profile sits between those of the Central Valley and the Eden Valley. The Shiraz from Gods Hill gets very ripe and plummy. Vines from Lyndoch are often used in blended wines because of their aromatic characteristics – Penfolds' RWT and St Henri, as well as Torbreck, for example, do this.

Vineyards

There are a number of vineyards that can be regarded as benchmarks for the area: the Grant Burge Filsell vineyard and the Dutschke St Jakobi vineyard are in the flatlands south of Lyndoch, right next to each other. The Filsell vineyard was planted in the 1920s on alluvial soil. The Filsell Shiraz is produced from it and the vineyard is also responsible for around 75 per cent of the fruit that goes into the ultra-premium Grant Burge Meshach wine. The Meshach has violet colour and perfectly expresses the elegant fruit flavours and the aromatic lift typical of this area.

The St Jakobi vineyard was a mixed agricultural undertaking until vines were planted here in the 1930s. Shiraz now covers one-third of the vineyard around the old St Jakobi church. The soil is dark grey loam over dark clay and the vineyard keeps the moisture well. Its best results occur in fairly dry years. Dutschke makes the St Jakobi Shiraz and the Oscar Semmler Reserve Shiraz from the special 2.5-hectare block in the middle of this vineyard. Both wines tend to show the black cherry and plum flavours that are characteristic of this area.

The Draycott vineyard, next to the cellar door of Burge Family Winemakers, consists of vines that are between 20 and 50 years of age. The vines produce dark fruit with lush flavours that are typical of the area.

Torbreck's Hillside vineyard sits on the northern slopes of Lyndoch on a property that is over 100 hectares. Some of it is planted with old Shiraz. The soil is red clay over limestone, with an ironstone layer in between, mixed with quartzite. It produces rich, dark fruit with lifted aromatics and fine tannins.

The Fromm vineyard is very small. Less than 1 hectare of Shiraz was planted in 1936 on its red–brown earth soil. The vineyard is on the border between this sub-region and Gomersal, and its elevation is only 150 m. It is one of the early-ripening Shiraz vineyards of the Barossa. Yalumba makes a single-vineyard wine from this site.

There are other excellent vineyards around Lyndoch, but their contribution is difficult to isolate because of the blending approaches many wineries take.

Williamstown

The second sub-region within the Southern Valley is the small wine-growing area of Williamstown. Williamstown is the southernmost region of the Barossa. It has the highest rainfall, with on average more than 700 mm of rain per year. It has a cooler climate than Lyndoch, not only because it is further south, but also because of its higher elevation: 250–300 m.

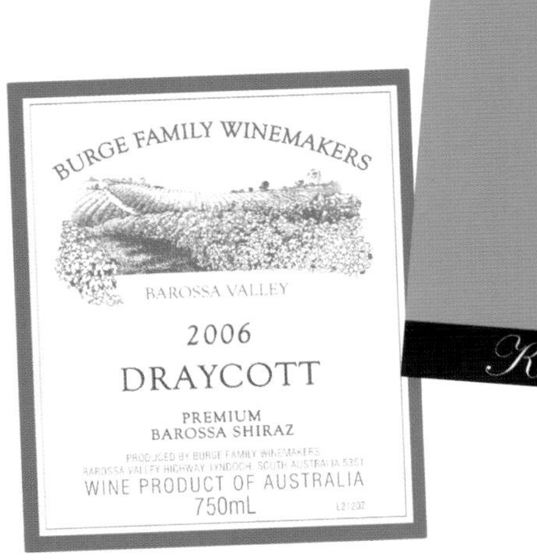

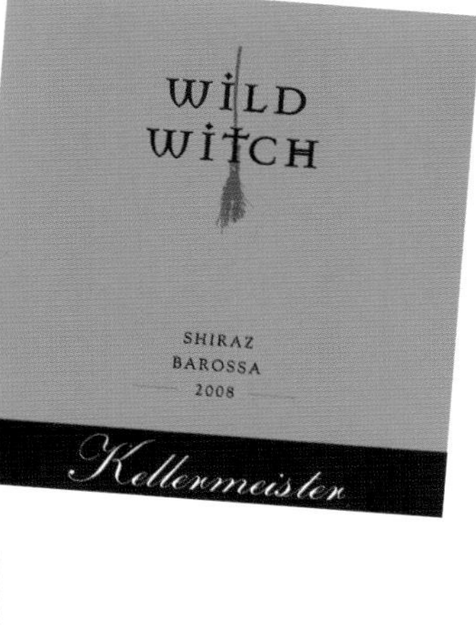

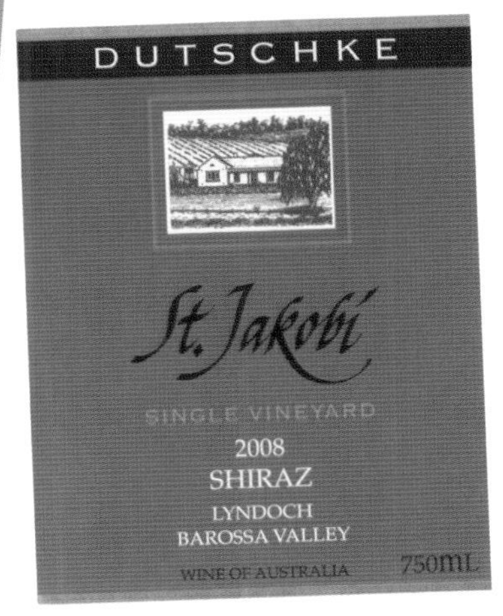

Williamstown

BAROSSA VALLEY

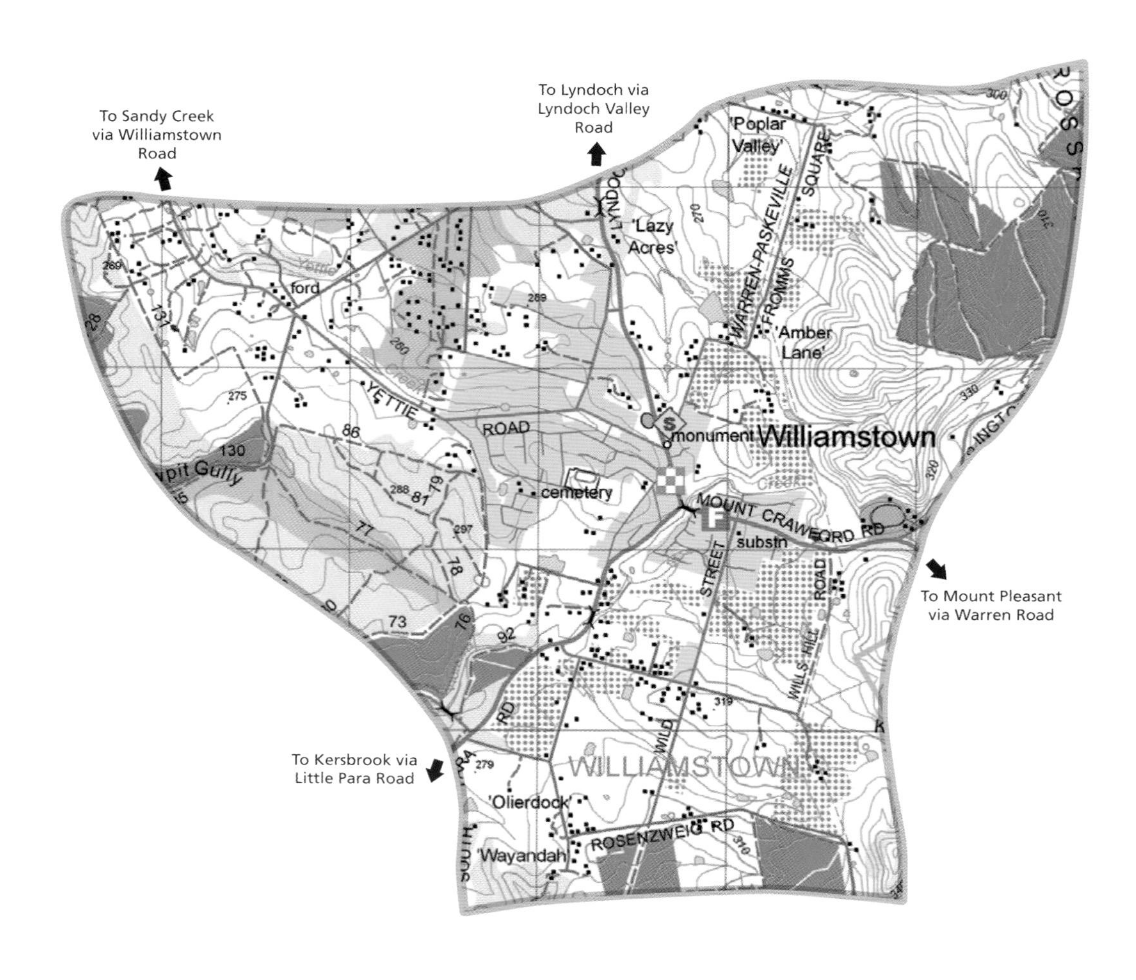
To Sandy Creek via Williamstown Road
To Lyndoch via Lyndoch Valley Road
'Poplar Valley'
'Lazy Acres'
WARREN-PASKEVILLE
SQUARE
FROMMS
'Amber Lane'
ford
YETTIE
ROAD
monument
Williamstown
cemetery
MOUNT CRAWFORD RD
substn
STREET
ROAD
WILLS HILL
To Mount Pleasant via Warren Road
WILD
WILLIAMSTOWN
To Kersbrook via Little Para Road
'Olierdock'
'Wayandah'
ROSENZWEIG RD
SOUTH

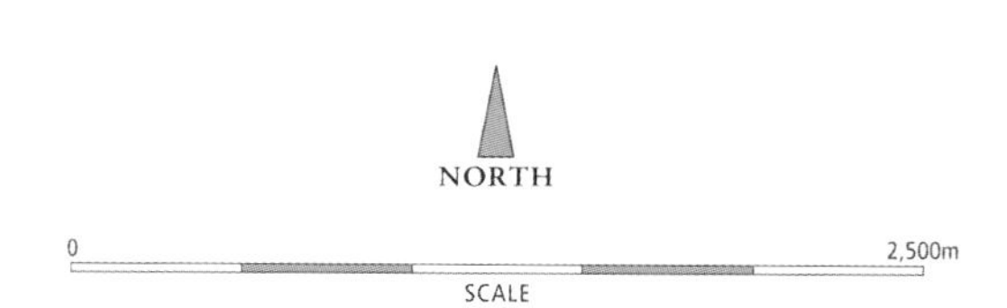
NORTH
0
2,500m
SCALE

Linfield Road vineyard

Williamstown is a pretty little hamlet, better known for its timber than its wine, as it is at the centre of the local pine forest industry. Mt Crawford forest and the whispering wall at the Barossa Reservoir are major attractions. Many of the original stone buildings from the mid 1800s are still in use.

A number of vineyards surround the town, the oldest Shiraz vineyard being a number of 100-year-old gnarled Shiraz vines next to the stone cottage of **Linfield Road Wines**, just south of Williamstown and close to the Adelaide Hills region. This property has been in the hands of the Wilson family since 1860. In 2002 they took the step from grapegrowers to winemakers. The main Shiraz is The Stubborn Patriarch Shiraz. The Edmund Major Reserve Shiraz is only produced in the best vintages. The soils of the vineyard are a mixture of red clay, slate, sandy loam and creek stones.

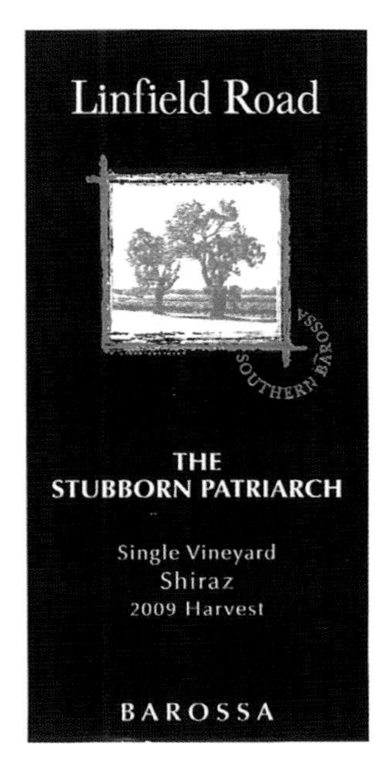

Red–brown earths and alluvial sands tend to be the main soils in Williamstown. The vines are harvested later than at Lyndoch, due to the higher elevation, which leads to a cooler climate and therefore longer ripening periods. The wines show red colour and have cherry flavours, often with earthy and eucalypt characteristics. They can be quite peppery, although generally less so than those from the Eden Valley. The finish of the wines is generally not very tannic.

The Leaders of the Barossa, Part 2

Prue and Stephen Henschke

The Henschkes are one of the most eminent winemaking families in the Barossa. Stephen, who has been the winemaker since 1979, and Prue are fifth generation Barossa people, and their children – in particular Johann, the eldest – are now taking on more responsibilities. Variable vintages in the 1990s, in particular for the Mount Edelstone wine, challenged their chances of maintaining the extremely high quality of two of Australia's most significant vineyards, Hill of Grace and Mount Edelstone (they are described in more detail in the Eden Valley chapter).

Stephen and Prue Henschke
[Photo: Dragan Radocaj]

However, the last few years leave no doubt that Prue and Stephen Henschke have managed to bring out the best in these vineyards by incorporating organic and biodynamic viticulture and by refining the wine styles. They have lifted the level of winemaking and wine quality in the Barossa, and for this alone they deserve to be included in this section. They also developed the cool climate vineyards at Lenswood and took the wines to the world market.

The Henschkes gain from the significant synergies of a close working team: Prue is the viticulturist and Stephen is the winemaker. They start with the objective of growing the best fruit possible – there is a particular focus on canopy management and soil health, and on minimal intervention, which is the philosophy throughout the winery. Stephen was one of the first winemakers to use screwcaps on premium red wines, and he introduced the Vino-Lok glass closure to Australia.

Prue and Stephen are very aware of the historical significance of the over 140 years of grapegrowing and winemaking they represent, and of the reputation they must maintain and build on. The meticulous program to look after the 'Grandfathers' (the original Shiraz vines of the Hill of Grace vineyard) and the careful replanting of cuttings to replace expired vines are testament to this.

Both Prue and Stephen have been very active in various Eden Valley and Barossa industry groups, and have chaired numerous committees. They are committed to the Barossa region and have been outstanding promoters of its wines. Stephen is a Baron of Barossa Grand Master, and together they won the prestigious Gourmet Traveller Winemaker of the Year award in 2006.

The Yalumba team

Most members of the group of leaders described in this book have been selected primarily for their contribution to winemaking. Robert Hill Smith has been singled out for his business skills. He is a fifth generation descendant of Samuel Smith, who founded Yalumba in 1849, and he and his brother Sam bought back the shares that had been sold to people outside the family in 1989.

Robert Hill Smith and Louisa Rose (above), and Brian Walsh (left).

Robert Hill Smith became Marketing Manager at 32 years of age, and two years later, in 1985, Managing Director. This was during a most difficult time for the region. He saw the consumer trend to table wines and moved Yalumba away from fortified wines and towards a portfolio of red and white wines. He also successfully introduced the 'premium' 2-litre cask and achieved market leadership in this category for Yalumba. Over time he has overhauled all the company's processes and improved their efficiency. Hill Smith, like the other winemaking leaders, is committed to the development of the region.

He focused the largest family company in the Barossa Region back onto the Barossa Valley and Eden Valley. Under his stewardship, Yalumba has become a leader in environmental management and the development of plant nurseries. He was instrumental in the establishment of an 'old vine charter', which categorises vines into different age groups – it has now been adopted by the Barossa Grape & Wine Association, the representative body of grapegrowers and winemakers in the region.

Hill Smith was also one of the driving forces behind the formation of 'Australia's First Families of Wine', which includes many well-established family companies, in 2009. Its objective is to lead the development of new ideas and new directions for Australian wine from a medium-sized company perspective in what has become, and may long be, a difficult period.

The recent success of Yalumba is not down to one man, however. It has been a team effort. But as always, some people stand out. The other two outstanding team members have been Brian Walsh and Louisa Rose.

Brian Walsh became Chief Winemaker in 1988, and 10 years later, as Director of Winemaking, became responsible for all aspects of production. This made him the bridge between production and marketing. He has provided the push for growth from the production end, and under his leadership wines such as The Octavius, The Virgilius and The Menzies have been created. He has also been a strong promoter of Yalumba wines and is a mentor for the current winemaking team.

Since becoming the Chief Winemaker in 2006, Louisa Rose has mostly been associated with the creation of white wines, in particular Viognier, and the introduction of The Virgilius at the top of the Yalumba Viognier pyramid. She is regarded highly by her peers, and as a show judge has won many prestigious awards. In recent years, she has introduced a small series of single-vineyard Shiraz and Grenache wines from the Barossa. The focus of these wines is on texture and drinkability; they do not follow the 'bigger is better' mantra of many others. She and Brian Walsh have led the efforts of the long-term 'Barossa Terroirs' project, which aims to identify *terroir* influences on Shiraz.

Employees do not leave Yalumba: the company's leadership team allows individuality and pioneering spirit to flourish within the company's business structure. If you have the chance to participate in a tasting of wines from Yalumba's legendary museum cellar, you are truly in luck.

John Duval

John Duval's reputation will always be inextricably linked to 1995, when the influential US *Wine Spectator* magazine crowned the 1990 Grange wine of the year. Duval was Chief Winemaker at Penfolds from 1986 to 2002, and during this period he achieved much more than that. He has been quite an innovator: for example, during his time Grange became a wine that could be consumed earlier, while still maintaining its rich and powerful style. He also created the new ultra premium RWT (Red Winemaking Trial) Shiraz, the first pure Barossa Shiraz, a wine that is matured in French oak, a departure from the traditional Penfolds style. Also, he won a number of significant international winemaking awards.

[Photo: Randy Larcombe]

Then it was all over. In 2003, Duval branched out on his own. He was excited to for the first time be entirely responsible for his wine. His knowledge of Barossa vineyards and growers was an ideal starting point. He focused on the red Rhône varieties, and Shiraz in particular. Interestingly, he is not trying to create a baby Grange. His trio of Plexus, Entity and Eligo are full flavoured and rich, but they also have freshness and finesse. This he has in common with Peter Schell of Spinifex. His colleagues regard him – and respect him – as a really good winemaker.

Because of his position in the industry, Duval travels overseas a lot. There he acts as an ambassador for the Barossa, pointing out what is unique about this winemaking area and its old vines.

Chris Ringland

When Robert O'Callaghan, in 1989, allowed his senior winemaker, Chris Ringland, to make some wine on the side – but not more than 2 tonnes, so that he would not get too distracted – he could not have known that this decision was the beginning of the Barossa Shiraz cult movement. Ringland decided to put his first Three Rivers vintage into high-quality new French oak hogsheads. The wine quality was good (the fruit came from the Gnadenfrei vineyard), so he decided to ask an outrageous price for it, given there was so little of it. Even today, not more than 100 cases are made in each vintage. American wine critic Robert Parker liked the style of this rich wine when he first tasted it in 1993: 'My God, this tastes like a pristine example of 1947 Cheval Blanc,' he said.[6] The rest is history, as they say: Parker has awarded Three Rivers points in the mid 90s, sometimes even 100.

[Photo: Toby Yapp]

This story is all the more remarkable because Ringland started out making aromatic white wine in New Zealand, where he was born. The Rockford training, however, converted him to substantial, rich red wines. Since 1998, the wine has been called Chris Ringland Shiraz, and since 1995 the fruit has come from the 100-year-old Stone Chimney Creek vineyard in Eden Valley, where he lives. The very ripe and thick-skinned fruit is matured in new French oak for over three years before release. Ringland's starting point when making this high-alcohol wine is to use fully flavoured grapes which can handle the alcohol. A lot of personal labour has gone into the vineyard management to achieve this. The wine has sometimes been described as a dry port. Despite its unique level of concentration, the wine can be quite balanced, and can display – besides dark fruit flavours – roasted meat, espresso, aniseed, charcoal and pepper.

6 In J.R. Guerra, 'The Talented Winemaker Chris Ringland', *The Valley Trends Magazine*, April 2011.

Chris Ringland Shiraz is now regarded as one of Australia's greatest wines.

There are other reasons why Chris Ringland has been included in this group of leaders. He is obsessed with winemaking and has been a consultant to many wineries in the Barossa, in particular Rockford, Greenock Creek, Turkey Flat and Hobbs. His influence cannot be overestimated. He is also involved in a couple of winemaking ventures in Spain. He has one of the most gruelling travelling schedules of any winemaker I have met.

Not everything has gone well. The higher volume R Wines, made in partnership with US importer Dan Philips, recently folded, after only a few years, but the Chris Ringland Shiraz story continues.

David Powell

David Powell started his winemaking career in the early 1990s at Rockford, under Robert O'Callaghan. He formed his own company, Torbreck Vintners, in 1994: he is Chief Winemaker, Managing Director and proprietor. Torbreck has grown quickly in size and reputation – current production is approximately 60,000 cases. It owns 80 hectares of vineyards, and purchases grapes from a similar amount of land from 40 independent growers. There has been considerable turbulence in the company's short corporate history, with Torbreck going into receivership in 2003. Powell stayed on and managed to buy back into the business. He now owns it, with Peter Kight of Quivira Vineyards in California. With the new winery in place as well, Torbreck's future looks much more secure now.

Powell adopted many Rockford philosophies, but he was also influenced by his wine experience overseas, and in particular in the south of France. His overseas experience did not just expose him to different winemaking techniques; it also taught him the uniqueness of the Barossa's dry-grown old vines, in particular Shiraz, Mourvèdre and Grenache. Torbreck wines are full-bodied, ripe and intense, but also balanced and elegant. Most Torbreck wines are a blend of grapes from different vineyards, and often of grapes of different varieties. While the type of wine is based on Rhône traditions, it is the *terroir* of the Barossa and the quality of the fruit that make these wines unique.

Two of Powell's other major contributions to the Barossa are mentioned often by other Barossa winemakers. The first is that he lifted 'brand Barossa' into the premium area – until the early 1990s this had not even been attempted. With the support of Robert Parker's reviews in the US, David managed, in a short period after forming Torbreck Vintners, to sell a number of mainly Shiraz-based wines for more than A$100 per bottle. This opened the door, and many others followed in his footsteps, finally able to build viable premium operations at relatively low volumes. In 2010, Torbreck set a new benchmark, releasing a single-vineyard wine, The Laird, at $700 per bottle, making it the most expensive current release in Australia.

The second contribution relates to relationships between winemakers and growers. Powell pays record prices for grapes in order to get the best raw material. He has hunted down old and neglected vineyards and nurtured them to benchmark status. In this way, income is passed on to a larger number of growers. Also, he emphasises source and location of the grapes more explicitly than others, and has published about his Barossa vineyards and their *terroir.* His winemaking philosophy is: 'Let the wine be.' For him, wine is made in the vineyard and the grower is the hero. The wines are true expressions of the place the grapes come from, whether that place is a single vineyard or – as is more often the case – a number of vineyards.

The Western Ridge

6

The Western Ridge constitutes the third section of the Barossa. In the southern part, it forms a major plateau, which rises from the Central Valley west of the North Para River. Further north it becomes an undulating area west of Stonewell Road, with the two major bowls of Marananga and Seppeltsfield. Many famous wineries are based in this section, from the historic Seppeltsfield to the relative newcomer Torbreck. The first sub-region is Gomersal, the second Marananga/Seppeltsfield. The third, further north, is Greenock.

Obvious differences in the *terroir* are that the Western Ridge has less rainfall than the Central Valley (380–450 mm as opposed to 450–550 mm per year), and is less fertile and more eroded. In this sub-region the grapes are exposed to a lot of heat from the sun in the afternoons.

Gomersal

Gomersal forms the most southern part of the Western Ridge. It is mainly a large plain, rising in the east and south from the North Para River. I have included She-Oak in the west, up to the border of the Barossa. The elevation is 220 m, except in the west, where it drops to 150 m.

Terroir

The sun is strong in this area, with temperatures higher than in the Central Valley and the nights usually warmer. The weather tends to come from the west, so the annual rainfall is less there, as more rain falls in the Central Valley because of the elevation to its east. On average it may be 400 mm per year. The area north of Rosedale is the hottest and has the earliest harvest.

There are three distinctive soil profiles. In the northern part, underlying Barossa ironstone creates an impermeable layer, which leads to an elevated water table. This results in cracking black soils during the summer months and water-logged soils in winter. These very black soils have traditionally not been regarded as suitable for wine production, but new vineyards have been planted here during the last 10–15 years.

Further south, the ironstone has cracked, and the soils now consist mainly of brown earth. This is where the original vines in Gomersal were planted. This same soil is in the new vine-growing areas in the southwest.

Finally, on the mid-western edge of the plateau, the surface soil consists of sandstone and is suitable for vineyards. However, while rain penetrates quickly,

Gomersal

BAROSSA VALLEY

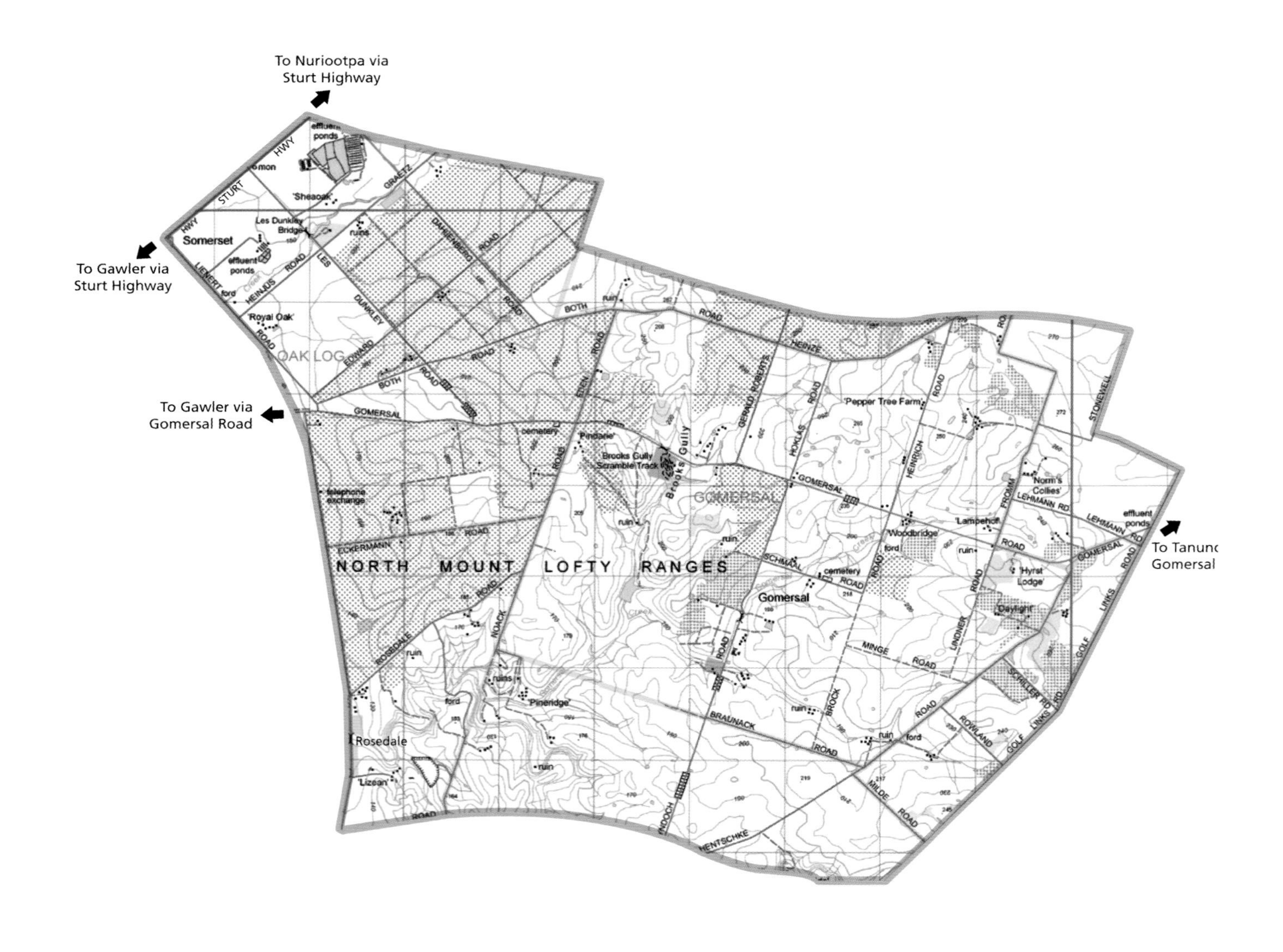

the water disappears just as easily. These sandy soils cool down in winter, and warm up quickly in spring, giving the vines an early start. The specifics of this depend on the depth of the sand and the sub-soils in each micro area.

Cellar door wineries

This part of the Barossa is less touristy. Significant parts are still farmland, and there are only two major cellar doors, far apart from each other.

Gomersal Wines is in the southern part of this sub-region. The wines come from the estate vineyard. The Gomersal Shiraz is quite big and dark, but offers peppery notes as well.

Further west along Gomersal Road, as the land gets hilly, is **Pindarie**, a winery started in 2005. The views are excellent and the cellar door has been established in one of the old farm buildings which the owners are restoring. The winery has a strong focus on sustainability. They are planting out a significant area to Shiraz, Tempranillo and Sangiovese. The Shiraz grapes grow on red–brown earth over limestone. Two Shirazes are made, the Pindarie Shiraz and the Black Hinge Reserve Shiraz. They are treated in American and French oak; 20 per cent new oak is used in order to enhance the tannin structure.

Significant areas, in particular north of Rosedale and towards She-Oak, have recently been planted by tax-advantaged investment schemes. I know little about the quality or destination of these grapes.

Tasting profile

It is perhaps too early to describe a typical flavour and structural profile of Gomersal Shiraz. There may be several, depending on the soil conditions. To date, Grenache has been the red grape of choice for Gomersal.

From several single-vineyard, as well as barrel, tastings, I would conclude that the Shirazes have a more concentrated and riper flavour than those from the Central Valley, but significantly less than those from Marananga/Seppeltsfield. Fruit flavours can be mocha-coated plum, but also more red fruit, in particular raspberry. The intensity of the tannins varies, but tends to be higher than in the Central Valley.

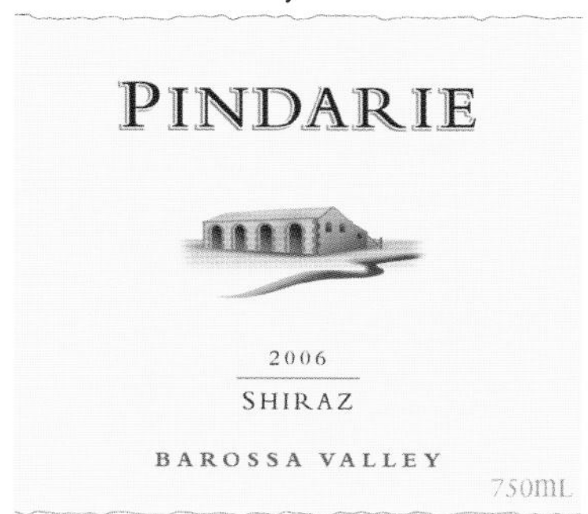

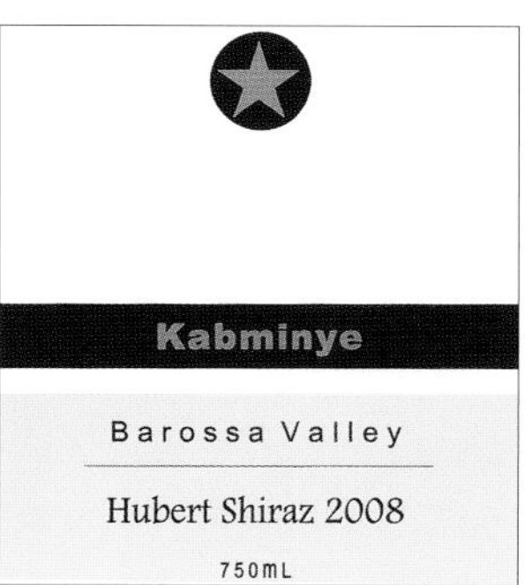

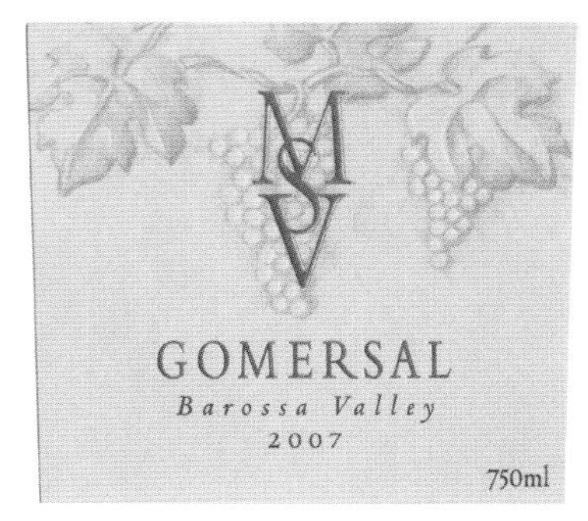

Marananga/Seppeltsfield

Marananga/Seppeltsfield is a small sub-region, not much more than 4 kilometres east/west and north/south, but it has received a lot of attention for a long time. It is home to the Seppeltsfield winery and complex – the best-known tourist destination in the Barossa (and one of the most visited in South Australia) – which has approximately 300,000 visitors per year. This is also the area where many 'big' Shirazes originate and where Robert Parker, the wine reviewer, initially focused his praise of Australian Shiraz. Its borders are Stonewell Road in the east, Heinze Road in the south, and the Sturt Highway in the north and west.

There are no major towns in this area, apart perhaps from the settlement of Marananga. It changed its name from Gnadenfrei in 1918 – a time when many German names were replaced by non-German names. The Gnadenfrei Lutheran Church remains as a reminder of the little town's heritage.

Terroir

Probably no other sub-region shows as much variation within a vineyard or between neighbouring vineyards as Marananga/Seppeltsfield. The bottom parts of the hills are usually covered by more soil than the top, which is also touched by cooler winds. There can be three weeks' difference in ripening time in vineyards next to each other, depending on temperature and aspect. Generally, budburst and ripening are later than in the Central Valley, but they happen fast. Vines can stress quite easily in this area because of the harsh conditions.

The soil in the Marananga/Seppeltsfield area is predominantly loamy and quite complex. The ironstone that is prominent in the Northern Barossa has been eroded here. There is now weathered rock on the hills and deep red–brown alluvium in the creeks, such as Greenock Creek. There are many different soils in the hills. Schist is exposed in the Seppeltsfield area, as well as grey and blue siltstone. Pink quartzite is visible at Marananga, in particular near Roennfeldt Road. As the quartzite is highly fractured, vine roots penetrate the red–brown earth beds between it. These stones may be over 500 million years old.

As a result of the differences in conditions between the top and the bottom of the hills, more cropping is usually required at the bottom end of down-sloping vine rows. Yield variability in this area is also much higher than in the Central Valley, as a result of the differences of soil fertility, soil–water relationships, weather conditions and vineyard management in different micro areas.

The Marananga/Seppeltsfield sub-region is an extreme vine-growing region by international standards. The grapes get incredibly ripe here, as a result of the reduced canopy, the generally eroded soil and the gentle slopes, which combine to ensure significant sunlight exposure. It is also possible that the quartz in the ground directs additional light to the vines, as suggested by biodynamic principles. There is limited moisture, and the rocks in the soil may contribute to temperatures remaining relatively high at night. The berries are small, and therefore have a high skin-to-juice ratio.

Major wineries

Almost at the centre of this sub-region lies **Seppeltsfield**, Australia's most iconic winery complex. It is more a village than just a cellar door. It was founded by Joseph Seppelt, a Polish migrant, who bought 63 hectares of land here in 1851 and initially set out to grow tobacco. The famous port store cellar, with a capacity of 9 million litres of fortified wine, was completed in 1878. It was built with the idea that it would store a barrel of port from every vintage for 100 years. In 1987, the first 100 Year Old Tawny Port was released. Seppeltsfield believes it is the only winery in the world which has now stored 130 years of wine without interruption. It is also the only winery which releases a 100-year-old wine every year. This port has become the signature wine of Seppeltsfield.

The winery grew quickly, and at the turn of the 20th century the Seppelt Winery was Australia's largest, with an annual output of 2 million litres. The winery's fortunes have not been so good during the last 25 years, after the family sold out of the business. The last ownership change occurred in 2007, when the business was bought by private investors with a link to the Kilikanoon winery. They introduced the Glenpara Table Wines, including the Dining Hall

The famous Seppelt port (fortified wine) store.

Marananga/Seppeltsfield

BAROSSA VALLEY

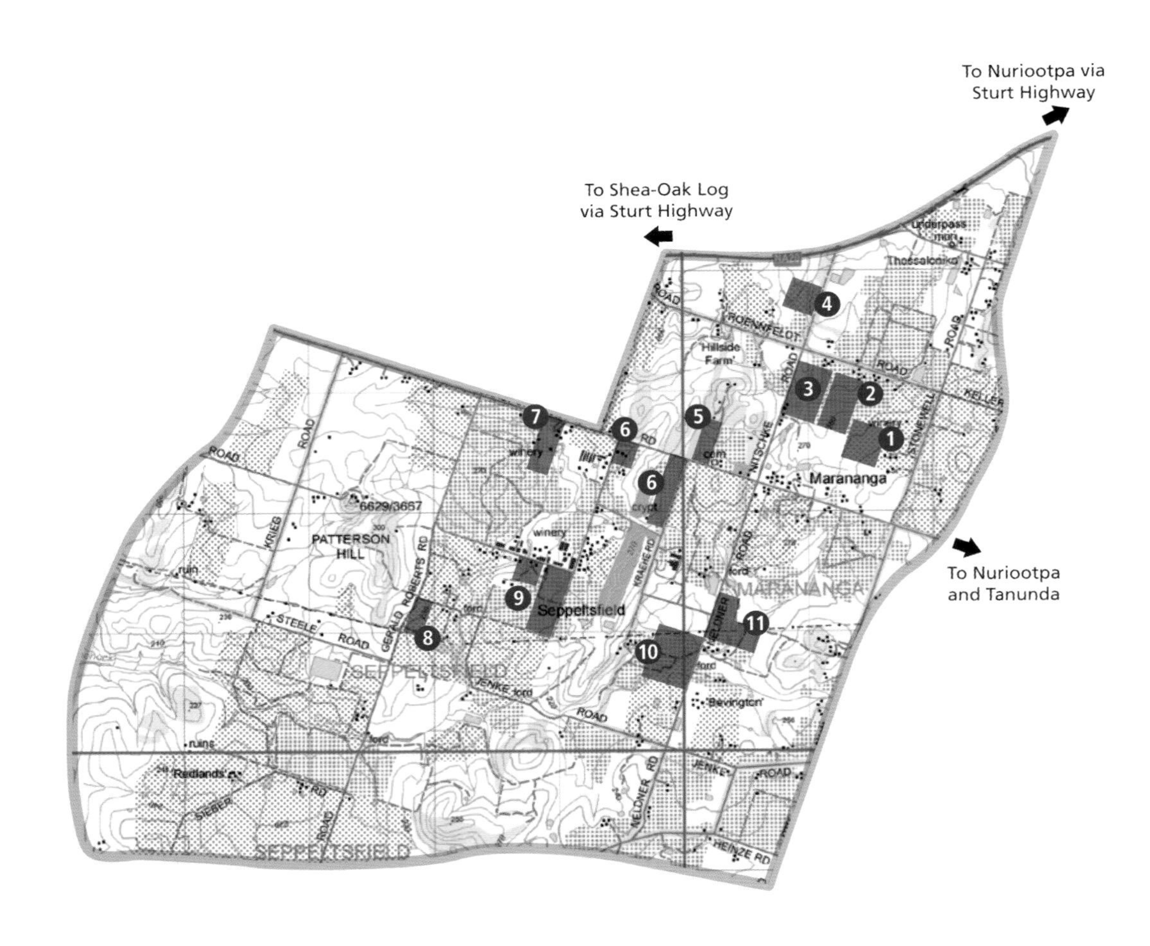

❶ Benchmark Vineyards

1. Torbreck, Descendent Vineyard
2. Zimmerman Vineyard
3. Greenock Creek, Roenfeldt Road
4. Powell Vineyard
5. Gnadenfrei Vineyard
6. Tscharke, Seppeltsfield Block and Home Block
7. Greenock Creek, Creek Block
8. Hentley Farm, Clos Otto
9. Seppeltsfield, B2 and B6 Block
10. Renshaw Vineyard
11. Neldner Vineyard

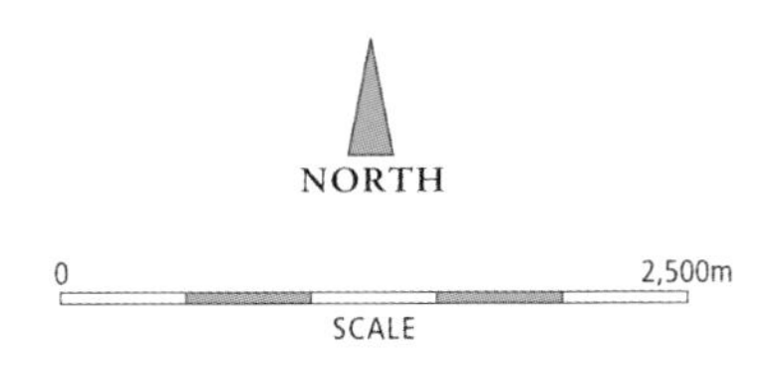

Shiraz. However, most of the premium fruit is still contracted to Treasury Wine Estates (previously Fosters). I anticipate that over time, and once label rights are regained, grapes from the old vineyards, in particular Grenache and Shiraz, will be used to make fine table wine. Interestingly, Penfolds, as part of Treasury Wine Estates, just released its first sub-regional wine from these vineyards, the Bin 150 Marananga Shiraz.

As one follows 'Palm Avenue', the road lined with picturesque date palm trees, to the east, one arrives at a large modern winery, **Barossa Valley Estate**. This winery was formed as a co-operative in 1984, when 80 growers banded together to make their mainly Shiraz grapes into wine rather than sell it off to others for little money or pulling out the grapes. Constellation Brands was a 50 per cent shareholder for many years, but the winery is now again 100 per cent owned by the co-operative. The flagship wine is the E&E Black Pepper Shiraz, which sources grapes from all over the Barossa, although predominantly from the Northern Barossa. Other premium Shiraz labels are the E&E Sparkling Shiraz and the Ebenezer Shiraz.

These Barossa Valley Estate wines are full-bodied and ripe, without being over the top. They display flavours of dark fruits with a twist of pepper and background oak. The fruit is concentrated and the flavour long, with a rich mouthfeel. Tannins are quite firm. These wines are made from the best fruit of the company's best growers, with the E&E range getting the top selection.

The philosophy of recently departed winemaker Stuart Bourne was to allow fruit derived from great vineyards to express its own unique characters in the

The well known palms near Seppeltsfield.

Vineyard near Two Hands

finished wine. This means working with great fruit and using only the bare minimum of inputs into the process; if you want to make great wine, this is, in Bourne's view, the only way. The winemaking technique is a mix of old-fashioned methodology with technology where it can help get consistency in results while preserving the integrity of the fruit – quite a tightrope walk at times, it would seem.

Next to it is the cellar door of **Two Hands Wines**. This initially virtual winery was formed in 1999 as a Shiraz specialist. A cellar door was opened in 2003, and a winery in 2004. It offers the Garden and Picture series, with Shiraz from many regions of Australia. All the wines are blended from fruit within a major region. Comparison tastings are available at the cellar door.

Bella's Garden, representing the Barossa, is probably the most popular. It usually features cherry and plum flavours, licorice, black tea and pepper. The grapes come from a number of vineyards, mainly from the Western Ridge and Northern Barossa. Following the purchase of Branson Coach House, the winery now offers two Shirazes from single vineyards: the Coach House Block Shiraz and Zippy's Block Shiraz.

Two Hands tends to pick relatively early to retain fruit purity and supple tannins for its full-bodied wines. Grape pressing is gentle and there is minimal intervention. Every parcel is kept separate until bottling. The winery has strong exports to the US, where it has won major accolades.

Further towards Stonewell Road lies **Heritage Wines**. This winery offers quite a contrast to the modern Two Hands, as Steve Hoff formed this winery in

1984 and set out, along with a few others, to preserve the heritage of the Barossa. It offers the traditional personal hospitality at the cellar door, with Steve behind the counter most days. If you bring a large container, he is happy to fill it with his port.

The Heritage Barossa Shiraz is made from 100 per cent low-yielding (5 tonnes or less per hectare) Marananga/Seppeltsfield fruit, most of it from the vineyard below the cellar door. The fruit is harvested based on flavour, typically at 14° to 14.5° Baumé. The Shiraz is aged for 18–24 months in new and old American oak barriques. The wine tends to be big and rich, with blackberry and chocolate flavours, and some oak noticeable.

Damien **Tscharke**, across the road, makes wines under two labels. The Glaymond label is reserved for Barossa's traditional varieties, and Tscharke is used for 'alternative' varieties such as Tempranillo, Montepulciano or Savagnin. The first wines were made in 2001, but grapegrowing has been in the hands of the Tscharkes for 6 generations.

The Glaymond Distinction Shiraz is made from low-yielding old vine vineyards nearby. The wine is ripe, with dark berry notes and minerality, and is enhanced by oak flavours. It is intended to show the flavours and textures that are derived from the vineyards as well as seasonal fluctuations.

Zippy's Block

Marananga Vineyard

Across on Roennfeldt Road is the new **Torbreck** winery. Dave Powell, who worked at Rockford in the early 1990s, founded Torbreck in 1994. The winery focuses on Rhône varieties, mainly red, but also white, and has taken the Rockford approach – acquiring the best fruit in the Barossa – to another level.

Torbreck believes that great wines rely on great *terroir* and viticulture, so their wines are made in the vineyard and then nurtured through to the bottle. Fruit yields are kept low through sourcing old vine material and through a rigorous winter pruning program done completely by hand. The grapes are harvested according to flavour and tannin ripeness, and the winemakers focus on developing a harmonious balance between power and elegance. The unique character of each of the vineyard parcels forms the cornerstone from which the individual blends are made. Gentle handling and crafting of the wines in the cellar and the intensely rich flavours that can be produced from dry-grown old vines are the hallmarks of Torbreck wines.

At Torbreck, old world winemaking techniques are used to make all wines, and they make minimal adjustments to the fruit in the winery. The hand-picked fruit is de-stemmed, not crushed, into small open fermenters, where it is left to sit for 24 hours. It is then inoculated and fermented for around seven days, during which time it is pumped over twice daily. The Shiraz ferments to around 24°C before the juice is cooled during pump-overs. They basket press the wine off skins at around 2–3 Baumé and allow the wine to finish the primary ferment in stainless steel tanks. It is then racked off gross lees and transferred to new and aged oak barriques. Malolactic fermentation is completed naturally and the wine is racked off malo lees and returned to barrel for between 18 and 36 months. The entire process is quite gentle and is designed to minimise maceration of skins and seeds. The best barrels are selected and the wines are bottled without fining or filtration.

The winery has been on a steep growth path, and it has not been without hiccoughs and corporate restructuring. Dave Powell's contribution to the Barossa is reported elsewhere in this book. Torbreck has been more explicit about the focus on vineyards than any other winery in the valley, and a lot of information about this can be found on its website. The major Shirazes are the flagship RunRig Shiraz Viognier (the blending approach is discussed in a separate chapter), the Descendant, a single-vineyard Shiraz/Viognier from the vineyard behind the winery, the Factor, a blended 100 per cent Barossa Shiraz, and the Struie, a blend of Barossa and Eden Valley fruit. The ultra-premium The Laird, vintage 2005, from the Gnadenfrei vineyard, was released for the first time in 2010. All wines are known for having masculinity, richness and ripeness, as well as elegant flavours.

Further west is **Greenock Creek Wines**. Michael and Annabelle Waugh established the winery in 1978, and released their first wines 10 years later. All five Shirazes are single-vineyard wines from vines up to 80 years old. The vineyards more or less surround the owners' home and the cellar door, which is closed for most of the year. Greenock Creek is a true estate winery, of which there are not many in the Barossa. The wines are known for their extremely high alcohol content, up to 18 per cent, and for their longevity. They enjoy cult status since Robert Parker showered them with praise and extremely high ratings. Michael Waugh is also a stonemason. He built the Rockford Winery buildings from ironstone in the 1980s.

Hentley Farm Wines in Seppeltsfield is a relatively new winery with great ambitions. The focus is on Shiraz which comes from vineyards surrounding the cellar door, itself an old shearing cottage. An attractive tasting room has been added. The vineyards include hill tops with little topsoil as well as alluvial soils of Greenock Creek, and have a range of aspects. The main Shirazes are The Beauty Barossa Valley Shiraz, The Beast Barossa Valley Shiraz, The Clos Otto, and the Barossa Valley Shiraz. The wines tend to be full-bodied with ripe tannins.

Sieber Wines is situated further south, bordering onto the Gomersal subregion. This property is still mainly a cropping and grazing farm. It added viticulture in 1998. It focuses, like so many others, on Shiraz. Special Release Shiraz and Ernest Shiraz are the two main labels.

Gnadenfrei vineyard

Tasting profile

The colour of the Marananga/Seppeltsfield wines is often black and inky, and the flavour profile is rich and intense, moving from blackberry to chocolate-coated plum. At the same time, the tannins tend to be silky or velvety. They can be powerful as well. There is a softness of mouthfeel, with a high pH and relatively low acids. Picking time has a huge influence on the final character of the wine.

Vineyards

There are, relatively speaking, more single-vineyard wines made in this sub-region than in any other in the Barossa, and a number of vineyards stand out as benchmarks.

The Gnadenfrei vineyard has an east and a west slope, and the soil and temperature conditions are quite different from the top to the bottom of the hill. While it is a single vineyard, it is this combination which provides the balance

Greenock Creek Seven Acre Block

and elegance in flavour and structure that the grapes from this vineyard deliver. The Shiraz vines of this 2-hectare property are 50 years old and the soil is dark heavy clay over red friable clay. This vineyard has delivered fruit to Rockford and Chris Ringland in the past and is currently the source of Torbreck's Laird.

The Descendant vineyard is the area behind the Torbreck winery. The 4 hectares are planted to Shiraz and other Rhône varieties. The soil is deep red clay over terra rossa. The vineyard already delivers wines with very intense flavours backed by silky tannins – typical of the Roennfeldt Road area – even though it is only 17 years old.

Greenock Creek's Roennfeldt 1.25-hectare Shiraz vineyard also sits on heavy red loam. Its 80-year-old vines often produce less than 2.5 tonnes of fruit per hectare and form the basis of an incredibly powerful Shiraz. The influence of the winemaker, however, is significant, and leads to a very alcoholic wine which masks, in my opinion, some of the character of the grapes.

David Powell's vineyard across the road grows 0.4 hectare of 100-year-old Shiraz vines. It produces very concentrated and dense grapes, which incorporate blackberry, violet and licorice flavours– a typical Roennfeldt Road old vineyard.

Greenock Creek's Creek Block vines are 50 years old and planted on fertile deep alluvial soils near the river bed. Again, thanks to the winemaking, the wine from this block is incredibly intense. The palate is thick and syrupy and tastes of blackberry liqueur and beef blood.

The 3-hectare Clos Otto Block at Hentleys Farm is a bit over 15 years old now. The soil is red clay loam, with only 30 cm of topsoil at the top of the

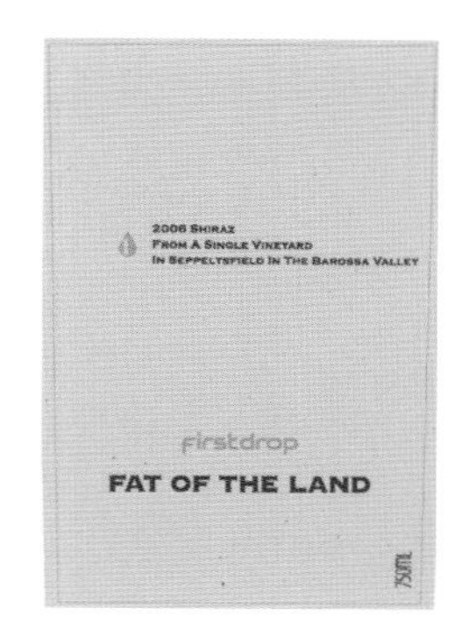

Penfold's Waltons Vineyard

hill – and 1 m at the bottom. The subsoil is red plastic clay which is difficult to penetrate but has great water-holding capacity. The characteristics of the wine produced from this block are intense blackberry and chocolate flavours balanced by ripe, supple tannins.

Tscharke's Seppeltsfield Block is situated next to 'Palm Avenue' at Seppeltsfield. The vines were planted between 1962 and 1965 on red clay over limestone. This soil has low fertility. The Glaymond Distinction Shiraz is made from these vines and vines from the small home block, planted on the other side of the hill in 1955.

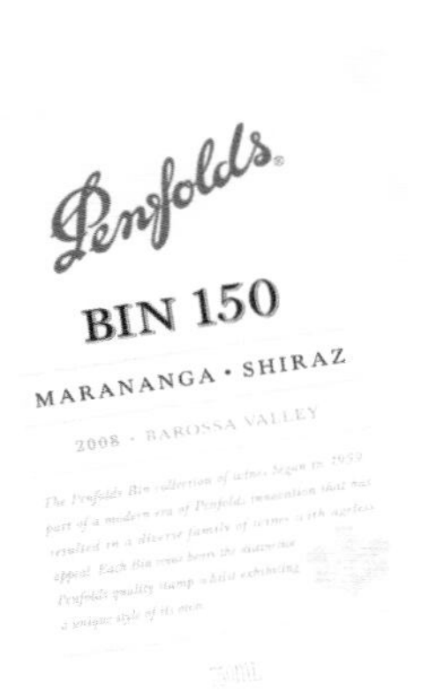

The outstanding Seppeltsfield Shiraz blocks are block B2, which is 1.4 hectares and was planted in 1974, and block B6, a larger 5.4-hectare block planted in 1994 on very deep red clay over shale bluestone. These vineyards sit on outstanding soil for the sub-region. Newer plantings across the ridge, in particular the C2 block, show great promise as well.

There are also two outstanding vineyards – and both are typical for the region – on Neldner Road: Torbreck's Renshaw vineyard and the Neldner vineyard. The old Shiraz grapes from these vineyards are inky black and produce very concentrated, low-yielding fruit.

Penfold's Waltons vineyard is also likely to become a benchmark vineyard for the region once the vines have aged more.

Greenock

Defining the Greenock sub-region presents a number of challenges. The Greenock Creek flows (if it flows at all) through Marananga/Seppeltsfield as well as Greenock, and in fact also through Moppa. The Greenock name is associated with the Greenock Creek Winery in Marananga/Seppeltsfield; the Kalleske Greenock single-vineyard wine, for example, is from south of the Sturt Highway as well.

I define the Greenock sub-region as starting on the ridge north of Radford Road. It is limited in the northeast by Spring Grove Road and then extends mainly west of the Kapunda Greenock Road. I have placed the hillier areas to

Wildflowers growing in the Barossa.

the northeast in the Moppa sub-region. This is perhaps a narrower definition than might be commonly associated with this area, but I believe it makes it more homogenous.

The town of Greenock is the centre of the Western Ridge. Its origins were as a resting and relay station for the hauling of copper from the Kapunda mine, but as a result of the fertile land surrounding it, it soon became a service centre for the farming community. Today, Greenock is a quiet and leafy town of fewer than 300 people. The local pub, which used to be a relay station for the mail coaches and groups hauling copper or farm produce, is the social centre of town. It is here that you can meet winemakers having a beer after a day's work – and you may also find rare bottles of wine behind the bar for purchase.

The Greenock Creek Tavern, and the red-brown soil profile of Greenock Creek.

Terroir

The Greenock sub-region is slightly higher than Marananga/Seppeltsfield, at an average elevation of 300 m. The dominant soil is friable red–brown earth. The colours vary from pale to bright. The soils can be quite rocky and shallow. There is also black loam in the vineyards towards Kapunda and iron-stone is present throughout. Soil conditions can vary considerably over a short distance. The vines are not very vigorous. Greenock is wetter than the very dry Northern Barossa, with about 500 mm of rain per year. Ripening is later than at Marananga/Seppeltsfield.

Major wineries

The two major cellar doors in this sub-region are Murray Street Vineyards and Kalleske. While Kalleske's cellar door is in the centre of Greenock, I will deal with this winery in the Moppa area, where its estate is located.

Murray Street Vineyards is a new winery owned by Andrew Seppelt, a descendant of the founders of Seppeltsfield, and partners. The first wine was

Murray Street winery

made in 2004. There is a *terroir* focus at this winery, which produces two single-vineyard Gomersal Shirazes, including the Sophia, from a special block there, and a single-vineyard Greenock Shiraz. They own a 24-hectare vineyard in Gomersal on rich red clay over slate at 200 m elevation, and a smaller one in Greenock, surrounding the winery, on brown and sandy loam at an elevation of 275 m. This vineyard includes 40-year-old vines. The Gomersal Shiraz fruit is quite exposed and ripens early, whereas the Greenock vineyard is more protected by hills and temperatures are somewhat lower, which means its fruit ripens later. In addition to the single-vineyard wines, a couple of blended Shirazes are produced. At this stage, most grapes are still sold to other winemakers.

Based on his experiences of making wine in France, the US and the Barossa, Andrew Seppelt wants to pick the best of new world and old world winemaking and vineyard management. Like many others, he believes that winemaking starts in the vineyard, and the vineyard needs to be in a great location. He employs minimal intervention in the vineyards. They will be picked over two or three times to ensure that the grapes are picked at optimal flavour and chemical balance. The Gomersal wine is matured in American oak, whereas the Greenock Shiraz is matured in French oak.

Where to list a virtual winery in a book about *terroir*? I have decided to list **First Drop Wines** in the Greenock sub-region, as most of its Shiraz seems to come from this area. First Drop Wines is Matt Gant and John Retsas. This very dynamic duo make their wines in the Adelaide Hills, and what they make is an ever-increasing array of unusual as well as traditional wines. They talk about their passion for life and fun and their desire to make 'kick-arse booze'. The

Greenock

BAROSSA VALLEY

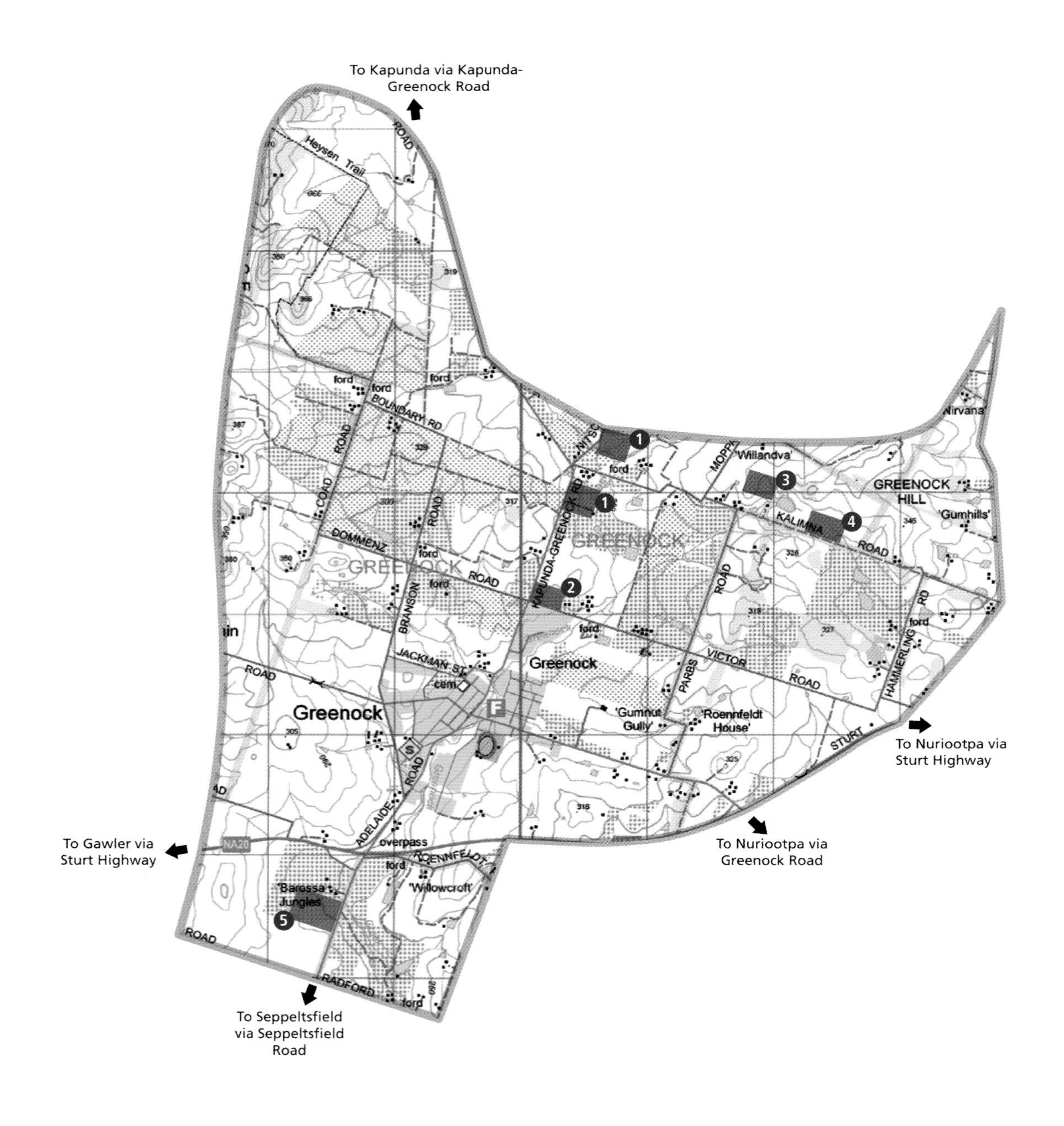

Benchmark Vineyards

1. Materne Vineyards
2. Victor Russell Vineyard
3. Geoff Brown Vineyard
4. Shawn Kalleske, East Block
5. Kalleske, Greenock Creek Vineyard

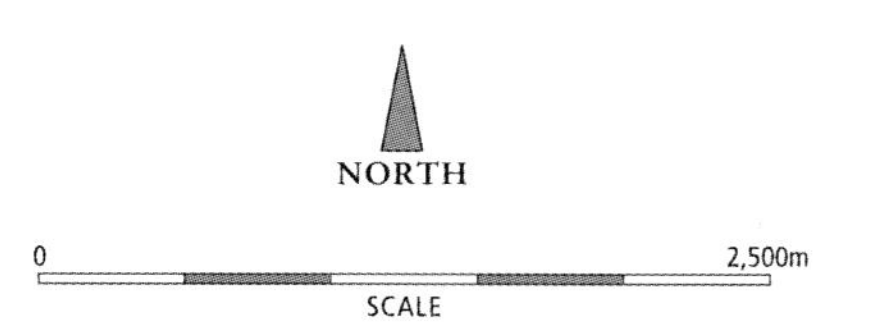

David Materne House Block

names and labels are often as in your face as the wine, but this should not deter: their wine, particularly the Shiraz, is high quality.

The two main blended Barossa Shirazes are Mother's Milk and 2%, the more serious wine. Its name refers to the splash of Tempranillo blended into this rich and fat Shiraz, which is sourced from Greenock, Marananga/Seppeltsfield and Koonunga/Ebenezer. The single-vineyard Shirazes are made in limited volumes, and called Fat of the Land. Gant and Retsas aim to show the range of contrasting styles that can be produced from the Barossa sub-regions. They describe the Seppeltsfield wine as rich, muscular and brooding, like Arnold Schwarzenegger; the Koonunga/Ebenezer wine as soft and textured, rich and big breasted, like Dolly Parton; and the Greenock wine as earthy, savoury and enigmatic, like Humphrey Bogart. The winemaking is similar across the range, but the oak treatment differs.

Tasting profile

The colour of Greenock Shiraz is very dark, and the flavours are of dark fruit, blackberries, black cherries and plum. There are earthy flavours as well, and often rich mocha or chocolate. The wines are bold and full-bodied, somewhat similar to those of Marananga/Seppeltsfield, but maybe not quite as big. They can be strongly structured, with tannins a bit more muscular and robust than those from the area's southern neighbour, but there are also examples of sweet wines with soft tannins and perfumed flavours, such as Turkish Delight.

Vineyards

The best known benchmark vineyard in Greenock, on the border with Moppa, is a couple of blocks of the Materne vineyard. It has bush Shiraz vines that are over 100 years old, grown on shallow sandy loam over yellow clay. These vines produce small dark berries, and wines with aromatic flavours and soft tannins.

The second major vineyard, south of the Sturt Highway, is the Kalleske Greenock vineyard. The soil is shallow sandy loam over deep red clay. The grapes produced here have intense blackberry flavours and taste of licorice and mocha as well. This is a typical expression of this sub-region.

A couple of newer vineyards, planted only in 1998 in the east of Greenock, show exceptional promise: the Shawn Kalleske East Block and the Geoff Brown vineyard.

Just north of the village is the old Russell Greenock Farm vineyard. Many varieties are grown in this vineyard, as was common in the early farming days; they include some old Shiraz.

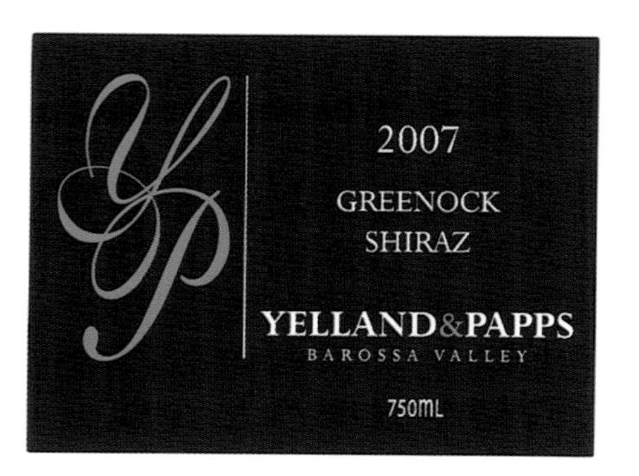

The Leaders of the Barossa, Part 3

The Kalleske Family

Tucked away in a quiet part of the Barossa, in the Moppa sub-region, lies the beautiful estate of the Kalleske family. Vineyards have been the focus of this mixed farming business since its beginnings in 1853. It remains a mixed farm today – the family breeds organic lambs and has a chaff mill – and is now increasingly looked after by seventh generation Tony, Troy and Kym, sons of John and Lorraine Kalleske, who are still involved in the business.

The core of the property is the 48 hectares of vineyard surrounding the source of Greenock Creek. Some grapes go back to 1875, and they are on average 50 years old. This vineyard, which has been tended organically from the beginning, and biodynamically more recently, has often delivered grapes for Penfolds Grange, and is regarded as one of the top vineyards in Australia.

Tony & Troy Kalleske
[Photo: Andy Ellis]

It took until 2002, however, for Troy and Tony Kalleske to make and brand their own wine. They started to bottle single-vineyard Shiraz as vineyard blocks came out of contract to other winemakers. All fruit used in the Kalleske labels is estate grown.

The Kalleske family is admired for a number of reasons. They are low key and unassuming. They had a tradition of nurturing the soil long before growing organically became fashionable. Kalleske serves as an example of how to successfully move from grapegrowing to winemaking. Today it is a true estate, with single-vineyard wines. And finally, they make very good wine. The winemaking is done with minimal intervention, so the flavour and structure are true to the place where the grapes are grown. The wines are lush and ripe, yet vibrant and fresh.

In a short period Kalleske wines have established themselves near the top of Barossa wines – this is the result of some unique *terroir*, sustainable vineyard management and a winemaking approach of minimal intervention and manipulation.

Peter Schell

Peter Schell is rumoured to have said: 'If you ever find me making a typical Barossa Shiraz, just shoot me.' He and his wife, Magali Gely, came to the Barossa in the 1990s from New Zealand – having spent five vintages making wine in the south of France. He established his winery, Spinifex, in 2001. The wines of Spinifex are certainly very different from those of most other Barossa wineries.

[Photo: John Krueger]

Peter Schell buys small parcels from everywhere in the Barossa, but often from the hillier areas of the Eastern Slopes and Eden Valley. He likes grapes grown in sandy soils, because they deliver aromatics and perfume. He also enjoys blending many varieties, in particular Mataro (Mourvèdre), Grenache, Shiraz and Cinsault, but also other French and Spanish varieties. He is highly respected for his intuitive method of winemaking.

Schell too talks about the primacy of the vineyard, so his blended wines are made only from grapes from what he considers the most suitable vineyards. He does blend many varieties and many areas, and those blends change from year to year. He experiments with new varieties too, and in 2008 made a straight Shiraz for the first time, and named it, rather humorously, Bête Noir.

Like all the winemakers featured in this book, Peter Schell likes old vines and low yields. However, he picks grapes early, which delivers a freshness to the wine and allows for lower alcohol levels. His philosophy and vision are unique: he aims to produce wines with intensity, but lightness. This makes his wines unlike other reds from the Barossa, but also unlike wines from southern France. Many winemakers have taken note, and I expect his influence to be significant.

Kym Teusner

I attended Kym Teusner's first winemaker dinner in Sydney a few years ago. He looked like a schoolboy, he was nervous as hell, and he mumbled some technical detail in what must have been his first ever speech. It was pretty disastrous. How far he has come!

Teusner started as an assistant winemaker at Torbreck and his first efforts under the Teusner label, the Joshua and the Avatar, have to be described as carbon copies of the Torbreck Juveniles and Steading GSM wines. He was fortunate, however, to draw on the Riebke vineyards in Koonunga/Ebenezer, and he priced the wines very competitively. Since then, he has gone from strength to strength. His wines are based mainly on fruit from unirrigated old vineyards in the Northern Barossa, most notably those owned by the Riebke brothers. Today, he makes some of the most honest and vibrant wines in the Barossa. His winery has grown from a standing start in 2002 to 15,000 cases in 2010. A major upgrade and expansion of the winery complex was taking place in 2011.

Kym Teusner is probably the only one in this group of leaders who has not distinguished himself by significant innovation. However, he is showing how hard work, ambition and attractive wines – wines that are based on rich, mature fruit, and that are well priced – can produce spectacular success in a difficult decade for makers of wine. As such, he sets an example for the Barossa.

Kym Teusner (right), with Leon Riebke in his vineyard.

7 The Northern Barossa

The fourth major section of the Barossa is often left off the tourist maps. There are no major wineries here and many roads are unsealed. This is truly agricultural land, but what land it is. Some of the best and most celebrated vineyards – not just in Australia, but worldwide – are in this sub-region. As a result, there has been significant new planting in the last 15 years. Over time, this area is likely to be as covered in vineyards as the Central Valley.

Three sub-regions make up this section. Moppa, marked by its undulating topography, lies to the northwest. The second major area is the flat sub-region of Koonunga/Ebenezer. The small area of Kalimna has been separated because of its special soil, and it forms the third sub-region.

Moppa

Terroir

Moppa is the area northeast of Greenock. It lies 50 m higher than Greenock, on average. As a result, it is usually not as warm, and it has quite cool breezes at times. Grape picking in this area takes place three to four weeks later than in the Central Valley.

The key soil characteristic, as in the rest of the Northern Barossa, is the hard layer of ironstone, which is up to 50 cm thick. The second major characteristic is the difference between land at the top of the hill and land at the bottom of the hill. At the top of the hill, the rock is eroded and the topsoil is shallow. There is a lot of sun exposure for the root system. At the bottom of the hill, there is deeper sand, often black. Moving up, the clay is mostly grey. Red–brown earth can also be found.

Major wineries

The main estate in Moppa is **Kalleske**. The Kalleske family is discussed in more detail in Chapter 6. The seventh generation of the family is now working on this 50-hectare site, which takes in the source of the Greenock Creek and surrounding hills. Kalleske is an organic estate winery, and started to sell wine under its own label only in 2002. As vineyards come out of contract, individual single-vineyard wines are made. The best known is the Greenock Shiraz, which comes from a single block at the southern end of Greenock. The main Shiraz wines from the estate are the Moppa, which includes a little Viognier and Petit

Verdot, the Johann Georg, which is made from fruit from the oldest vineyard, a dry-grown vineyard planted in 1875, and the Eduard Shiraz. The last two come from benchmark vineyards, and they demonstrate the depth of flavour and balanced structure that can be achieved in this sub-region. Kalleske has opened a cellar door in the centre of Greenock.

The Kalleskes committed to working their land sustainably long before organic products and biodynamic farming methods started to become popular. They believe it is the way nature intended the process to function, and they believe that farming this way makes better wine. Healthy soil is the prime basis for healthy vines, expressive grapes and quality wines. Farming the vineyard organically and biodynamically results in more individualistic wines, wines that are true to their site. For the vines to capture the unique characteristics of soil and climate with their roots and their leaves, the soil must be alive and healthy and the leaves must be free of chemicals. Kalleske continues the organic and biodynamic theme from the vineyard into the winery, relying on natural yeasts for the primary fermentation and natural malolactic bacteria for the malolactic fermentation. Kalleske does not use added tannins or fining agents and the wines are naturally clarified through gravity racking, without filtration.

The premium Kalleske Shirazes tend to be black in colour with bright, yet intense aromatics. The wines are full-bodied, quite ripe, with complex flavours of black cherry, blackberry, chocolate and spice. The tannins are fine and silky and the wines have great length.

Westlake Vineyards is based on a Moppa vineyard, which was in the hands of one arm of the Kalleske family for a couple of generations. It was given to Suz and Darren Westlake. After having supplied many leading wineries with grapes for some years, they started making their own wine in 2003 and produced their first wine under their own label in 2005.

The winemaking philosophy is to keep it simple, to select the best parcels of fruit and to highlight the different soil types in the vineyard. The wine should have an ability to age. The winemaking style is typical of small-volume, premium producers: open fermenters, hand plunging, no tannin additions, extended lees contact, extended maceration, basket and bag pressing and a limited amount of new oak.

Westlake offers three Shirazes: the Eleazar, the Albert's Block and the 717 Convicts The Warden Shiraz. The wines all share intense dark berry fruit, chocolate, aniseed and coffee, which are typical of this *terroir*. They show a strong tannin structure. The Moppa block, where these wines come from, has good topsoil and very little stone. It is dominated by heavy red clay.

Shawn Kalleske, a 6th-generation grapegrower and distant relative of the other Kalleskes, started **Laughing Jack Wines** in 1999. His aim is to showcase

Moppa

BAROSSA VALLEY

❶ Benchmark Vineyards

1. Kalleske, Johann Georg Vineyard
2. Kalleske, Eduard Vineyard
3. Robert Burdon Vineyard
4. John Nitschke Vineyard

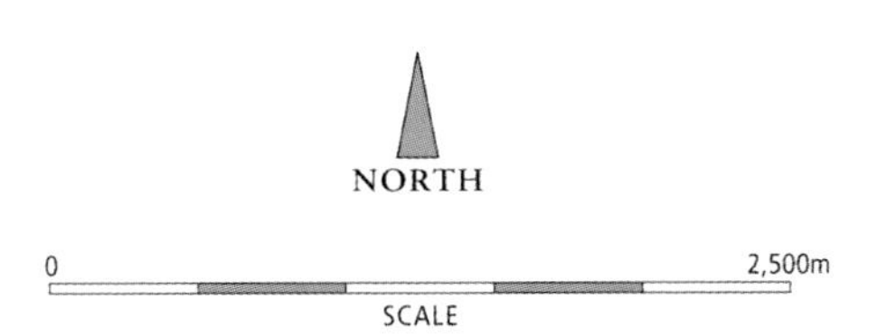

the wines' *terroir* while also enhancing varietal character and vintage footprint. The wine style is concentrated and balanced, with great fruit density and lifted aromatics; the wines are approachable when young, but also have the ability to age. They are matured in hogsheads for 18–30 months, depending on wine style and vintage.

Jack's Shiraz is a soft, supple and aromatic wine with chocolate, blackberry and plum fruit characters. The sandy soils of the Hill block and the Home block in Moppa are responsible for these characteristics. The Greenock Shiraz shows rich black fruits and a stronger tannin structure. It comes from the part of the Hill block where the topsoil is shallow and the red clay is close to the surface, and from the East block from the Greenock sub-region, which has deep red clay with a lot of ironstone. This is also the source of the very limited and powerful Limited Two Shiraz.

Head Wines is the result of Alex Head's move into making his own wine in 2006. He has access to two vineyards, one at 370 m in Moppa and the other next to Stonewell Road. The Shirazes are called The Brunette and The Blonde, a reference to the Côte Brune and Côte Blonde in Côte-Rôtie, which he admires as a wine district.

His objective is to make a delicate, fresher, less alcoholic and extracted Shiraz. The wines are treated in French oak barrels of different sizes, with new oak limited to 25 per cent.

Tasting profile

Shiraz from Moppa is quite concentrated, and that from small berries at higher elevation can be quite tannic. It also shows spice and savoury characteristics. In the valleys, and where there is sand over clay, the berries are bigger, and there are thus more aromatics and softer tannins. The flavours are often blueberry and red fruits and currants. There are rich chocolate flavours from the old vines.

Wines from Moppa are vibrant and quite balanced, with lower pH than those from Greenock (say, 3.4 versus 3.8), and higher natural acid levels (6.5 to 7 g/L).

Vineyards

The best known vineyards in the area are the Kalleske vineyards, which form the base of the Johann Georg and Eduard Shiraz. They are situated on the alluvial flats of Greenock Creek. The vineyard for the Johann Georg was planted in 1875 and is very low yielding. The grapes for the Eduard come from vines that are 50–100 years old.

The two other benchmark vineyards are the John Nitschke and Robert Burdon vineyards. They both sit on sandy topsoils and have delivered fruit to ultra premium wines of larger producers.

There are a number of premium vineyards in this area, but their fruit tends to be blended with grapes from other areas. Shawn Kalleske makes wine from Moppa vineyards, as do Rockford, Alex Head and a couple of other winemakers, but there are no cellar doors in Moppa.

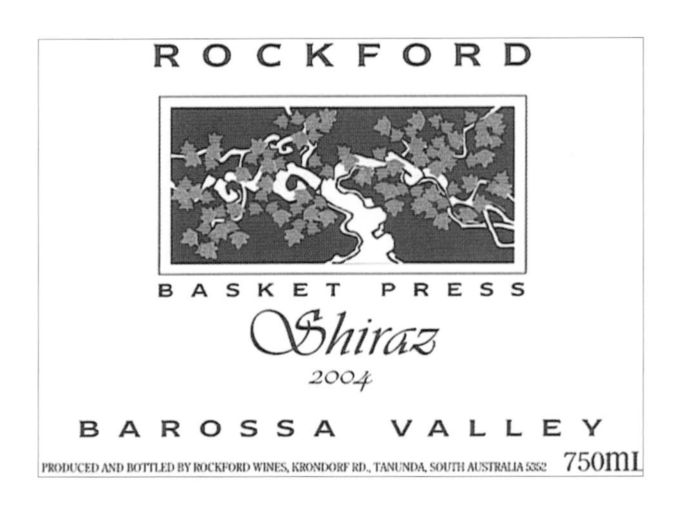

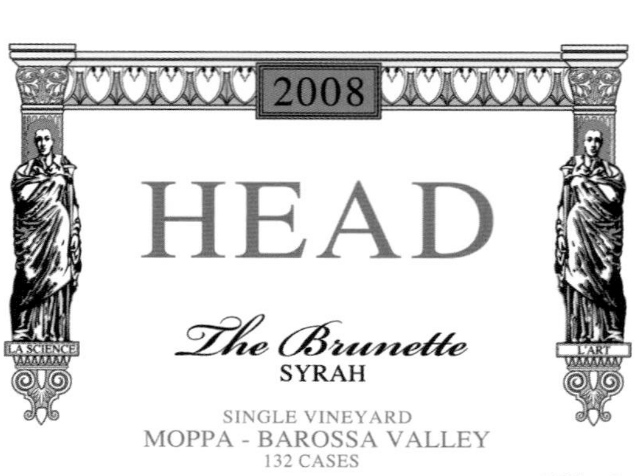

Koonunga/Ebenezer

Koonunga/Ebenezer is the largest sub-region of the Northern Barossa. It is to the east of Moppa and starts north of Pipeline Road. There is no township in this area. Major roads crossing this area are Research Road, Belvedere Road, Diagonal Road and Kapunda–Truro Road. This region is basically flat, at 270–280 m elevation.

This sub-region had its beginnings, like the rest of the Barossa, in the 1840s. But in contrast to the Central Valley, the pastoralists here amassed large properties. As no town developed, the Lutheran Church, built in 1859, became the centre of the farming community. In the 1860s, more settlers arrived, smaller blocks were allocated, and activity switched to other agricultural pursuits. Minerals, in particular copper, and marble were found in the area and created excitement; during these years, the first vineyards were also established.

Terroir

The Koonunga/Ebenezer sub-region is the warmest in the Barossa, 2° to 4° C warmer than Nuriootpa. The climate is more continental, with dry air and cool nights, which results in later ripening than in the Central Valley. This region is also the driest, with 10–15 per cent less rain than the Upper Central Flats.

The key feature of this region is its soil. This area is home to the unique Barossa ironstone, which is in fact cemented quartz gravels, enriched with iron. This iron-rich cap rock is very old, maybe 200 million years. Below this now crumbling surface is a layer of fine, mostly red clays. This soil has excellent water-holding ability. There are also calcium deposits on most exposed land, in particular in the eastern part of this region. The red–brown surface soil tends to be quite thin in most parts and the subsoils are quite hard. That hard subsoil, plus the high pH and sodium concentration, tend to limit root density and growth. The red–brown clay is deeper in the eastern part of Koonunga/Ebenezer, which means the land is more fertile there.

Vineyards

There are no wineries in this area, although Marcus Schulz is selling some wine under his own Schulz Vignerons label, as Benjamin Shiraz and Marcus Old Shiraz. The 38-hectare vineyard carries grapes that are between 20 and 70 years old. Grapes from the core bush block are sold to Penfolds and Torbreck. Close to Kalimna, the soil is sandy loam over red and yellow clay. This is one of the benchmark vineyards of the region.

Then there is Scholz Estate, with a 40-hectare vineyard which the couple only started to plant in 1998. It demonstrates that young vines can aspire to set a benchmark as well. The soils are loamy over red clays, typical for this area.

Koonunga–Ebenezer

BAROSSA VALLEY

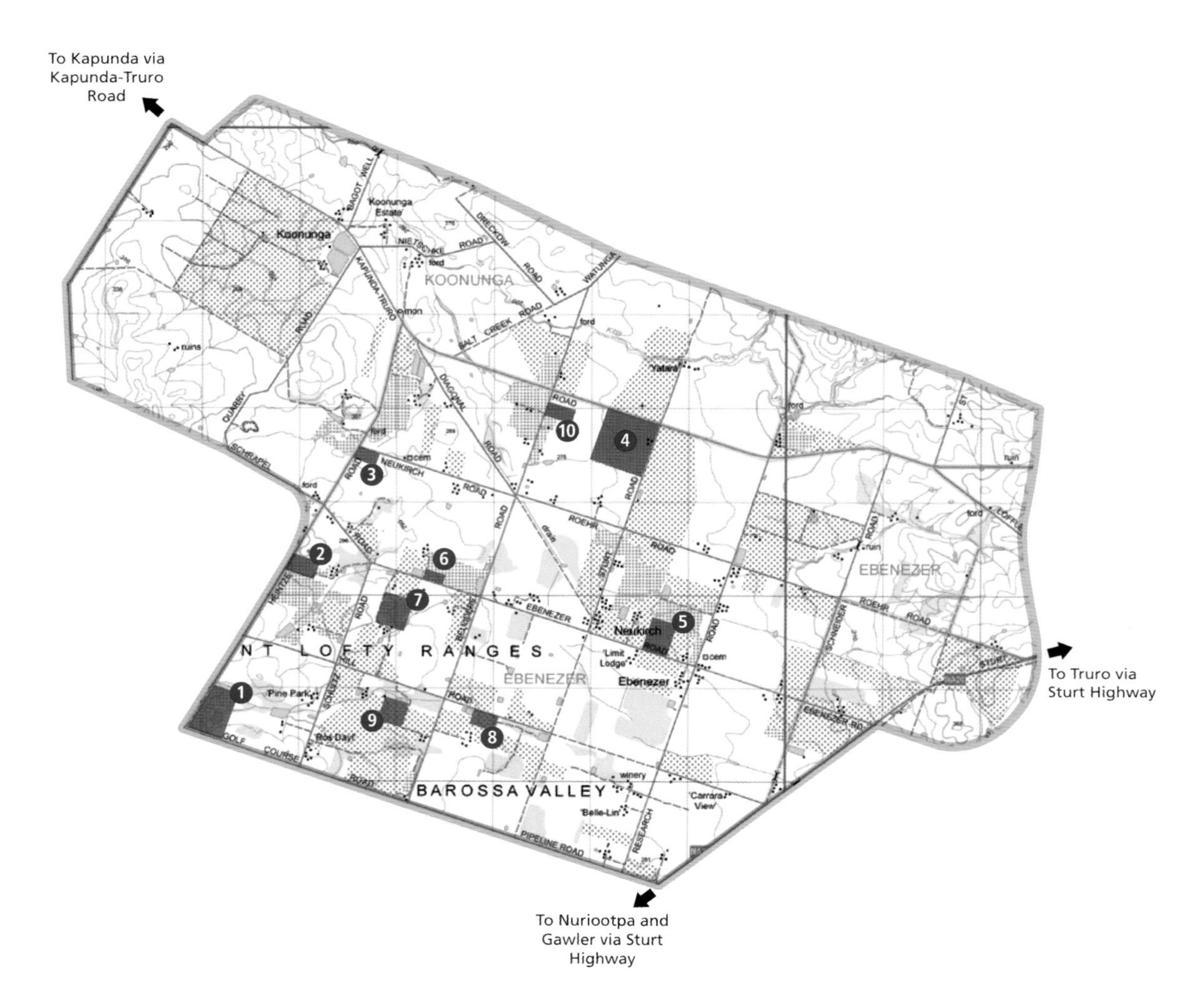

❶ Benchmark Vineyards

1. Penfolds, Block 42
2. P. Kleinig, Section 1
3. Heintze Vineyard
4. Penfolds, Koonunga Hill
5. A. Hoffman Vineyards
6. Schrapel Vineyard
7. Riepke, Home Block
8. Frank Gallash, Fig Block
9. Schulz, Bush Block
10. J. Nietschke Vineyard

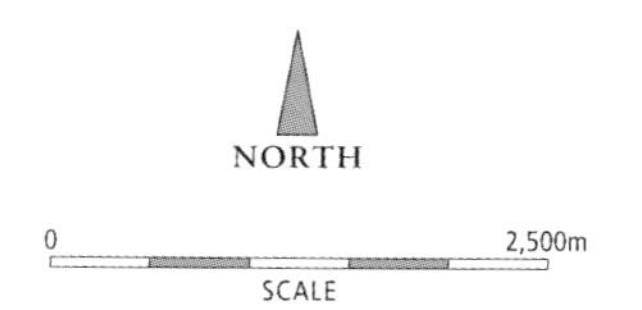

Ancient vines in the Schrapel 1885 vineyard.

Anthony Scholz is pushing the branding of his vineyard, maybe taking a few leaves out of the books of some leading growers in the Napa Valley. His grapes are found in the premium wines of John Duval and First Drop Wines. St Hallett is a major customer as well.

At the northern end is Penfolds' Koonunga Hill. This 63-hectare vineyard was planted in 1973. Its soil is red–brown earth over heavy clays. It has an average rainfall of 500 mm per year. It inspired the Koonunga Hill brand, although the grapes sourced for that wine come from many areas. The low-yielding vineyard is a consistent contributor to Grange and a producer of core fruit for Penfolds' St Henri, Bin 389 and Bin 28.

Another outstanding vineyard is the Dimchurch vineyard, in particular the Old Block and the Home Block. This 48-hectare vineyard also sits on red–brown earth over red clay, but with a layer of calcium. The Shiraz vines are up to 100 years old. Adrian Hoffmann, owner and vineyard manager, is a major supplier to Ben Glaetzer and Torbreck, and so to their premium wines – Amon-Ra, RunRig and Factor – and has supplied pretty much all Barossa royalty over the years.

Another special vineyard is the Schrapel family's 1885 vineyard. This vineyard has been the most prominent supplier to Peter Lehmann's Stonewell Shiraz. Peter Lehmann occasionally fashions a single-vineyard wine from it, The 1885 Shiraz.

The oldest block of the Riebke brothers, which has 130-year-old Shiraz vines, is another vineyard that can be seen as a benchmark for the region. It supplies Teusner Wines.

There are a number of other vineyards which have often supplied fruit to leading Barossa wineries, in particular to Penfolds. Some are very well looked after, some appear somewhat neglected. Marked on the map are Peter Kleinig's Section 1, the Heintze vineyard, the J. Nietschke vineyard, and Frank Gallash's Fig Block.

Probably the most famous of all Barossa Valley vineyards, Penfolds' Block 42, is above the golf course, in the southwest. The vines date from 1888, and are thought to be the oldest Cabernet Sauvignon vines in the world today.

Tasting profile

Shiraz wines from the best vineyards in the Koonunga/Ebenezer region have outstanding ageability. The berries are usually small, leading to a high skin-to-juice ratio. This leads to dark, big, ripe and very tannic wines with good length and persistence. The tannins of wines from the western part are often coarse-grained rather than the silky expression typical for Marananga/Seppeltsfield. The flavours include fruitcake, chocolate, mocha and licorice. The wines can be more generous and opulent in areas with deeper and more fertile soil, in particular in the eastern part. In this case, the flavours are more of big red fruits and Christmas cake.

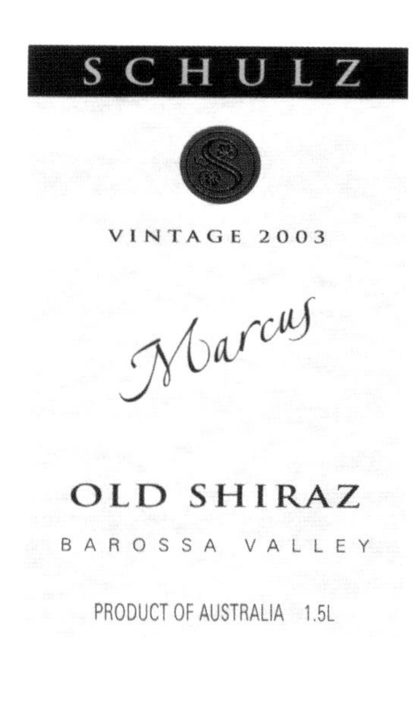

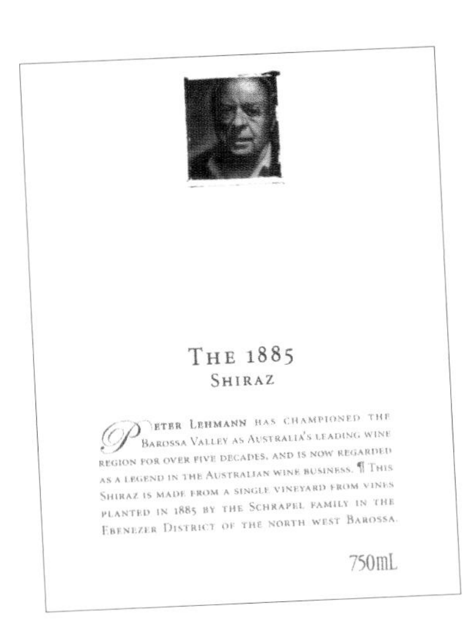

Kalimna

Terroir

Kalimna is a small area north of Nuriootpa and to the east of the golf course, and it is special because of its soil. It is quite different from the rest of the ironstone-dominated Northern Barossa. Its unique feature is the extensive sand cover, which is not found elsewhere in the Barossa to this extent. It can be up to 1 m deep, and more on the east–west trending sand dunes. There are many layers of sand, interspersed by very thin red dust layers. The origin of this sand is still being debated, but what is clear is that exceptional wine can be made from grapes grown here.

Vineyards

The key vineyard in this area is Penfolds' Kalimna vineyard. It is up to 340 m in elevation. Annual average rainfall is 500 mm, similar to Koonunga. The soil is deep sand and sandy loam over heavy red–brown clay. Penfolds purchased this vineyard in 1945, but grapegrowing goes back to 1888. Today, the vineyard covers 150 hectares of vines, mostly Shiraz and Cabernet Sauvignon. The average age of the vines is more than 50 years. The Kalimna vineyard is an exceptional Shiraz vineyard, and it is called the 'Mother vineyard' of Grange nowadays. It also provides core fruit for RWT, St Henri, Bin 28 and many special releases, such as Bin 60A, regarded by many critics as one of the best Australian wines ever made. It is regarded as the leading Penfolds vineyard in cool years; the Koonunga vineyard, which is on red–brown earth, is favoured in warm years.

Kalimna

BAROSSA VALLEY

1 Benchmark Vineyards

1. Penfolds, Kalimna Vineyard

Tasting profile

The sandy soil delivers a different profile from the rest of the Northern Barossa. Shiraz flavours from Kalimna are aromatic and lifted. Wines still have a big mid-palate and great length, but the fruit is not as concentrated, spices are soft, and the wine is not as tannic. The purity of the fruit stands out. Kalimna wines can vary quite a bit from year to year, depending on the rainfall. Wines from drier years are more highly regarded.

The Blending of RunRig

Blending has been a major feature of the Australian wine industry. This is also true for the Barossa. Historically, blending was carried out as an insurance policy, a 'just in case' option: just in case a grape variety does not ripen properly or a region gets wiped out by natural disaster, for example. And who knows, if the European rules were not as tight as they are, maybe a lot more blending would occur there as well. Certainly many wine companies in Italy have given up their precious appellation for some freedom to blend.

There is, however, a second, a more proactive approach to blending, in which different sub-regional characteristics are actively sought to create a more complex wine – in terms of flavour or tannin, say. I call this 'strategic blending'.

A number of the premium Barossa Shirazes, including Penfolds' Grange, St

Hallett's Old Block and Barossa Valley Estate's E&E Black Pepper Shiraz, are blended from fruit from many vineyards. It is impossible for an outsider to say if this represents a strategic approach or simply a hunt for premium grapes. I will use one less complex example, Torbreck's flagship Shiraz, RunRig, to illustrate how strategic blending works.

The objective for RunRig is that it should be an opulent, rich, concentrated wine.

While I have received information about grape sourcing from the winery, the interpretation of this blending that follows is entirely mine.

The Shiraz component of the RunRig fruit – there is also a small percentage of Viognier in the wine – is sourced from eight vineyards. Given the objective, my guess is that the majority of the fruit comes from the Northern Barossa and the Western Ridge.

The fruit from the Schulz and Hoffmann vineyards in the Koonunga/ Ebenezer region gives the wine fruit concentration, power and a strong backbone. The Powell vineyard and the Renshaw vineyard in Marananga/Seppeltsfield add to the dark colour and ripeness, but also soften the tannins somewhat and make them silkier. The Moppa fruit component, from the Moppa and Materne vineyards, sits between those two areas, as far as structure is concerned. Then there are contributions from the Hillside Vineyard in Lyndoch and the Philippou Vineyard in the Lower Central Flats. Before these are added, the wine is very powerful and meaty, so fruit from these two vineyards will add aromatics and a slightly lifted feel to the wine (reinforced by the Viognier), in order to give it better drinkability and add complexity to the favour profile.

In this way, fruit from vineyards and sub-regions is selected to create a wine that carries within it 'the best of all worlds'. It is the winemaker's skill that then ensures that all the required characteristics are reflected in the final product. RunRig is an example of the way in which the different characteristics of the Barossa sub-regions can contribute very specifically to blended wines.

Eden Valley

8

Eden Valley forms the second GI (geographical indication) of the Barossa, next to the Barossa Valley. It is of similar size. The highest ridge of the Barossa Ranges, between the two GIs, constitutes a relatively easily defined border.

Eden Valley's topography and climate are quite different from those of the Barossa. The elevation of Eden Valley is between 325 m and 570 m, basically close to twice the elevation of the Barossa. Rainfall is 20–40 per cent higher than in the Barossa Valley, and falls mostly during the winter months. However, it reduces towards the east. Temperatures are cooler, particularly in the areas of higher elevation. The soils are different as well. There is a lot of variation, but Eden Valley is dominated by rocky podzolic soils with low surface soil fertility, with red–brown earth only dominant in the north.

Eden Valley stretches from Truro in the north to almost Mount Pleasant in the south. It is not a valley as such, but a landscape of rolling hills, rocky outcrops and river valleys. Water management is very different from the Barossa, where the water table is close to the surface in many areas. In Eden Valley, water is captured and managed through dams.

The *terroir* is probably even more varied than in the Barossa Valley, due to these topographic and soil differences. However, I will not make an attempt here to define sub-regions of Eden Valley, although the high-altitude area, High Eden, has been defined officially as a sub-region. The main reason for this is that the discussion of the Barossa sub-regions has been based on Shiraz, but the main grape of Eden Valley is Riesling, certainly by reputation. Therefore Riesling should probably be the defining wine style. There is now slightly more Shiraz than Riesling grown in Eden Valley, but its tonnage is only 15 per cent of what is harvested in the Barossa Valley. This means the density of Shiraz vineyards is quite low, which would make generalisations difficult.

Having said this, there are some very old and important Shiraz vineyards in Eden Valley. Also, it appears that Eden Valley Shiraz is becoming quite sought after by many Barossa winemakers as average temperatures rise and Barossa Shiraz faces the criticism of being too big and too ripe. I will therefore attempt a general description of the conditions of Eden Valley and its parts without defining sub-regions.

The settlement of Eden Valley is closely linked to that of the Barossa and its early pioneers. George French Angas's son-in-law, Henry Evans, built a grand homestead near Angaston in 1843 and from 1850 established a pastoral property, vineyard and winery at Evandale, near Keyneton. Wealthy English families established country estates nearby, and took up grazing, farming and horticultural activities. Angaston was surveyed in 1841 and developed as the early centre.

The township of Eden Valley had its beginnings in the early 1850s, when wealthy William Lillecrapp bought land near the current town. The land was subdivided in the early 1860s and a village was established. It is believed it got its name from the word 'Eden' being found carved in a tree. Today, Eden Valley has a population of a little over 400.

A number of wineries that are still well known were established in the early years. The earliest and best known was Pewsey Vale. Its first vineyard was planted in 1847 by Joseph Gilbert, at high altitude. Its wines were well regarded in Australia and overseas, and established an early reputation for Eden Valley. Samuel Smith first planted the Yalumba vineyard in 1849. Johann Christian Henschke planted his first vines near Keyneton in 1865, and sold his first wine in 1868.

The name 'Eden Valley' began to be used in the 1950s as a description of this area. It was made better known by John Vickery, who used it on the label of his Leo Buring Rieslings. Yalumba continued its focus on the area with the purchases of Pewsey Vale and Heggies.

Major wineries

There are two outstanding cellar doors in Eden Valley, both in the northern part. **Yalumba** is regarded as the oldest family-owned winery in Australia. It was founded in 1849 by Samuel Smith, a British migrant. He purchased 12 hectares of land near Angaston. The magnificent cellar buildings, in particular the two-storey clocktower building, were constructed in the early 1900s. Today, Yalumba is a serious producer of most mainstream varieties, in particular Riesling, Viognier, Cabernet Sauvignon and Shiraz. It owns vineyards and has access to growers in many parts of South Australia. It is also the only winery in Australia with its own cooperage. Its best known Barossa Shiraz is The Octavius, a blend from Barossa vines, all of which are at least 35 years old. In the last five years, Yalumba has introduced premium single-vineyard Shirazes from Light Pass, Lyndoch and Eden Valley.

The winemaking philosophy is to make wines that are balanced, that are representative of the places they come from and that are savoury and dry. Wines are tasted often during the winemaking process. Yalumba tries to interfere as little as possible, but reaction to taste includes cap management during fermentation, storage time and type of oak.

The flagship Octavius Shiraz usually shows deep red–purple colour. The aroma should be opulent, full of plums, coffee and cedar, with a crunch of dark

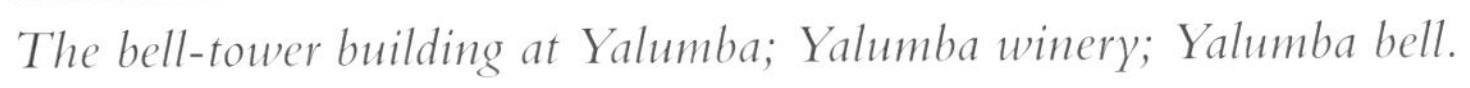
The bell-tower building at Yalumba; Yalumba winery; Yalumba bell.

Keyneton–Moculta

EDEN VALLEY

Benchmark Vineyards

1. Henschke, Hill of Grace
2. Henschke, Mt Edelstone
3. Fechner Vineyard

berries and a burst of olive and aniseed. On the palate there is a lot of spicy fruit. It is not a massively concentrated wine, but it is full-bodied and has a melt of oozing tannins to finish. The wine is typically a blend of Upper Central Flats and Eden Valley vineyards.

For more detail about the contribution of the Yalumba team, see Chapter 5.

Henschke is located between Keyneton and Moculta. It is a small cellar door, but what history it encapsulates! Henschke's first wine sales were in 1868, after Johann Christian Henschke had purchased and planted a small property in North Rhine, later named Keyneton, in 1862. The Henschke motto is 'exceptional wines from outstanding vineyards'. Their two flagship single-vineyard wines, Hill of Grace and Mount Edelstone, reflect and exemplify this philosophy. The connection to the Hill of Grace vineyard was through strong family ties with the original owners, the Stanitzkis. It was bought by Paul Gotthard Henschke in 1891 and remains in the hands of the wider Henschke family. The Mount Edelstone vineyard was purchased in 1974 – Henschke had made the Mount Edelstone Shiraz, from grapes planted in 1912 by Ronald Angas, since 1952.

Sustainability is a major theme for this winery, and Prue Henschke is regarded as one of the pre-eminent representatives of organic and biodynamic vineyard principles. Henschke's wine philosophy is to ensure that the soil preparation maximises the natural fruit flavours. The fruit is tasted in the vineyard

Hill of Grace vineyard

pre-harvest: the aim is to achieve pure fruit flavours and mature tannins. It is then hand-picked, and only the best fruit is brought to the winery and allowed to ferment – in traditional open-top fermenters using gentle, long, cool submerged-cap fermentation. No fining is used, and minimal filtration is applied, so that flavour, structure and wine complexity can be maintained.

The Hill of Grace tends to be dark crimson in colour. It has a complex nose of blackberries and blueberries, exotic spices, vanilla and cedar. The palate is sweet and fleshy, with great depth and texture, excellent length and intensity and fine, velvety tannins. The Mount Edelstone is darker, almost black, in colour, with a fragrant bouquet of plum, blackberry, anise, pepper and vanilla. The mouthfeel is often bigger than that of Hill of Grace. The wine has elegant and firm mature tannins contributing to a long finish.

The Henschkes have introduced a number of new wines during the last 10 years. At the premium end for Shiraz, they are Hill of Roses, from young vines of the Hill of Grace vineyard, and Tappa Pass Shiraz, from old vines supplied by growers from Tappa Pass in Eden Valley and Light Pass. Today there is also a large range of white wines, Shiraz blends and Cabernet Sauvignon from Eden Valley and the Adelaide Hills. So-called VIP tours of the vineyards and winery, including tasting of the premium wines, can be arranged.

Another interesting place to visit in Angaston is the **Taste Eden Valley Wine Centre**, which showcases a number of smaller wineries, among them Eden Hall, Poonawatta, Radford and Torzi Matthews. **Smallfry Wines** has opened a cellar door nearby in an old bank building, selling wines from High Eden and Vine Vale, where they manage a block of mixed grapes in the traditional Barossa style.

East of Angaston is the cellar door of a recent 'upstart', **Thorn Clarke**. David Clarke, a trained geologist, purchased a large amount of vineyard land in the late 1980s: in the Barossa Valley and Eden Valley, and also outside these GIs, further north in St Kitts. The company's philosophy is to blend fruit from four different vineyards. The premium William Randell Shiraz is made from Barossa Valley fruit, and the higher volume, well-priced Shotfire Shiraz is sourced from all of Thorn Clarke's Shiraz vineyards.

Terroir

As mentioned above, the *terroir* in Eden Valley is quite varied. The northern part, east of Angaston and south of Moculta, can be seen as a transition area between the Barossa Valley and Eden Valley. The soil here is red–brown earth over limestone, similar to the Northern Barossa. This area is 360–400 m in elevation, higher than the Central Valley of the Barossa, but lower than Eden Valley further south, and therefore warmer than that area. The two best known

'Grandfather' vines in Henschke's Hill of Grace vineyard.

Henschke vineyards, Hill of Grace and Mount Edelstone, are in this northern part. The Shirazes from this area tend to be full-bodied, but more aromatic and elegant than most Barossa Shirazes.

Vineyards

The Hill of Grace vineyard – or Gnadenberg, as it was originally known – is probably Australia's most celebrated vineyard. The Grandfathers, as the oldest block is known, was first planted in around 1860 by Nicolas Stanitzki, Stephen's ancestor on his grandmother's side. These vines, according to Prue and Stephen Henschke, were from the original vines brought out by the German Lutheran settlers in the 1840s. These vines, from pre-phylloxera material, survive today on their own roots. The Henschke family bought the vineyard in 1891. Currently, 8 hectares are planted with vines – 4 hectares with Shiraz, the rest Riesling, Semillon and Mourvèdre – at an elevation of 400 m. The average rainfall is 520 mm per year. The vineyard is actually situated in a valley and has a variety of soil types. The Grandfathers sits on sandy and silty loam over deep red clay. In the eastern part of the vineyard, the red clay is thinner and overlain by a thin layer of gravel wash. As a result, the ripening process varies across the vineyard. Very old schisty rocks make up the bedrock. These variations are responsible for the finished wine's complexity and require very detailed vineyard management. There is a very sophisticated and specific replanting program, based on vine selection, designed to ensure the long-term viability of the vineyard.

The Hill of Grace grapes tend to deliver a medium-weight wine in the plum and blackberry spectrum, with meaty, smoky and leathery flavours. A lot of

Henschke's Mt Edelstone vineyard

the complexity comes from the exotic spices. The flavours are often supple and velvety, with fine-grained, powdery tannins.

The Mount Edelstone vineyard is southwest of Hill of Grace and 4 kilometres west of the Henschke cellar door. It has an altitude of 400 m, and 600 mm rainfall per year. It is a 16-hectare vineyard, planted to Shiraz only. This was unusual when it was planted by the Angas family in 1912: most owners hedged their bets by planting a number of varieties in each vineyard. The material came from Joseph Gilbert's nursery at Pewsey Vale, from pre-phylloxera vines imported by James Busby into New South Wales. The first single-vineyard Mount Edelstone wine was made by Cyril Henschke in 1952. He finally purchased the vineyard in 1974. The soils of the vineyard are fine sandy loams over deep gravelly red clays overlying laminated siltstones. A scientific replanting program that is similar to the one used at Hill of Grace is being used at Mount Edelstone.

Mount Edelstone grapes have slightly more palate weight than Hill of Grace grapes. The fruit flavours are blackberry, mulberry and plum. There are also minty flavours, as well as cinnamon and pepper, and a strong chocolate and mocha component. The tannins tend to be firm and mature.

Another benchmark vineyard in the area is the Fechner vineyard; a number of its blocks are on Moculta Road. The grapes form the core of the St Hallett's Old Block Shiraz. They also go into Henschke's Tappa Pass Shiraz, the Kellermeister Wild Witch and the Dan Standish Borne Belline.

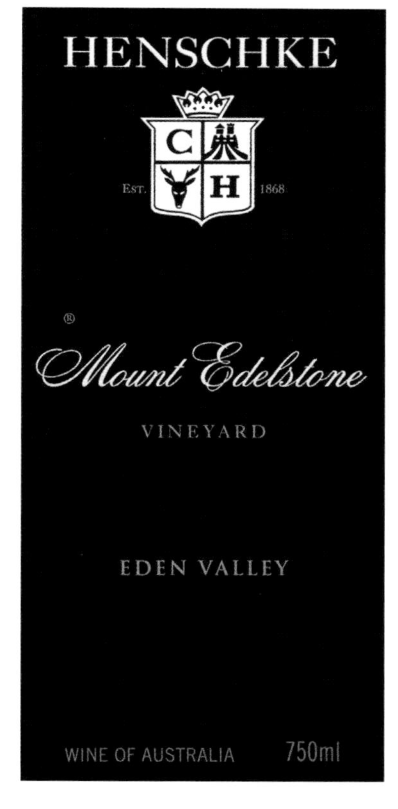

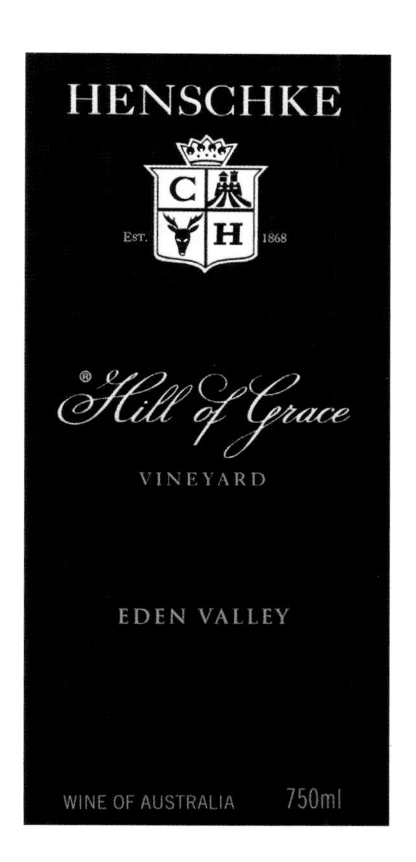

Mt McKenzie

The next major area is Mt McKenzie, on the eastern side of Eden Valley and with Flaxman Valley in the west. This area is more elevated and has lower rainfall than other parts of the valley. The soils vary a lot, but sandy loam dominates. Schisty rock is found there as well. There are some good quality Shiraz vineyards here at the end of Sawpit Gully Road, sometimes referred to as the 'Golden Triangle'. Core fruit for Yalumba's Octavius used to come from these vineyards. Today, the Knight vineyard is used for Torbreck's Gask Shiraz. It is a 2-hectare vineyard at 325 m elevation, planted with 45-year-old vines. The soil is sandy

Sawpit Gully Road vineyard

loam over sandstone, and is quite infertile. The cooler climate and soil characteristics are responsible for elegant and lifted flavours. The wine shows spice as well. The other two vineyards are owned by Severin and Trevor Hartsch. Some of the grapes find their way into the premium Shirazes of John Duval. The Torzi Matthews single-vineyard Frost Dodger Shiraz comes from here too. The vines are grown on clay loam over a schist rock bed. The palate weight of wines from this area is not as big as it is in those from further north.

Wineries/Vineyards

In the Flaxman Valley is **Chris Ringland**'s 1.5-hectare vineyard**,** south of Stone Chimney Creek Road. These 100-year-old vines are on poor soil and have very low cropping levels, sometimes below 2.5 tonnes per hectare.

Chris Ringland's Stone Chimney Creek vineyard

Chris Ringland started his side project, Three Rivers, in 1989 while at Rockford, and with the blessing of Robert O'Callaghan production was capped at 2 tonnes. This meant the wine would be hard to get: Three Rivers soon became Australia's most prominent cult wine. Since 1995, this wine, now called Chris Ringland Barossa Shiraz, has come from the Stoney Chimney Creek Road vineyard. Its colour is inky and it is incredibly concentrated, with dark blackberry fruit, espresso and barbecue flavours.

Chris has done a lot of work in the vineyard to improve the quality and health of the vines. His winemaking philosophy is that the vineyard, the grapes and the season should tell you what wine the grapes will become, rather than the other way round. His winemaking technique is heavily influenced by his experience at Rockford. He has a focus on the flavour and structure of the grapes, and so aims for gentle vinification and not over-extracting the tannins. He pays a lot of attention to every part of the process and keeps detailed records.

Hobbs owns the southern part of that same vineyard and also produces big and alcoholic Shirazes: one blended with Viognier, one straight variety wine, and one made in an Amarone style.

A relatively new winery in the Flaxman's Valley is **Radford Wines**. Ben Radford moved from the Barossa to South Africa, and made wine there for over

Mt McKenzie–Flaxman Valley

EDEN VALLEY

To Angaston via
Eden Valley Road
To Tanunda via
Mengler Hill Road
To
Keyneton
via Eden
Valley
Road
To Eden Valley via
Angaston Road
SOUTH MOUNT LOFTY RANGES
FTY RANGES
RANGE
KAISERSTUHL
CONSERVATION PARK
FLAXMAN VALLEY
MOUNT MCKENZIE
BOUNDARY
PEGGY
HILL
Mount McKenzie
Collingrove
Landers Corner
1
2
3
4
5
1 Benchmark Vineyards
1. Trevor Hartsch
2. Severin Vineyard
3. Knight Vineyard
4. Poonawatta Vineyard
5. Chris Ringland, Stone Chimney Creek Vineyard
NORTH
0
2,500m
SCALE

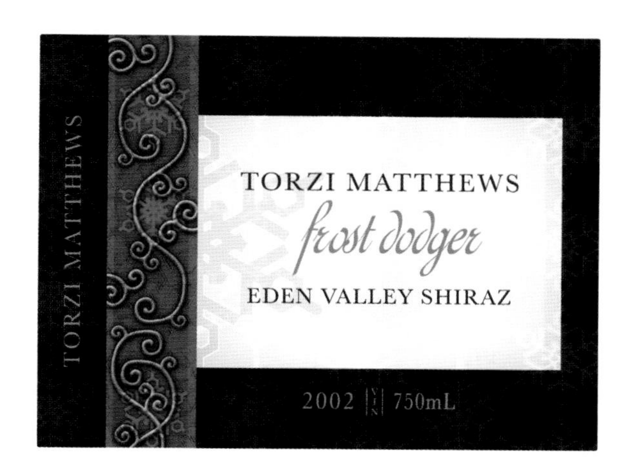

10 years. He and Gill, whom he met there, decided to move back to Eden Valley, and they made their first wines there in 2003.

While their focus is on Riesling, they also make the well-textured Eden Valley Shiraz. The fruit for the Shiraz is taken from a small vineyard planted in 2000, with some small batches from neighbours in the area. The vineyard is biodynamically managed, and the Radfords believe this will enhance the expression of their *terroir* in the wine. Their grey–brown duplex and yellow podzolic skeletal soils with quartz ridges are typical of the area.

Key features of the winemaking process include the use of predominantly wild ferments and French hogsheads, about 80 per cent old. The resulting Shiraz is elegant, with good length and berry and aromatic spice flavours, typical for Eden Valley.

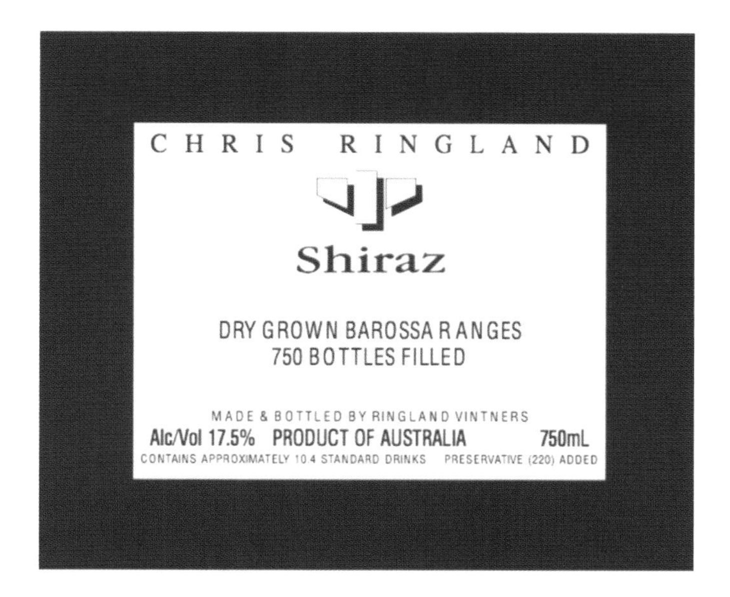

Craneford and Eden Valley

These areas are further south; skeletal soils dominate. Vines on this tough surface seem to deliver very aromatic and elegant wines. The wines often have a redcurrant and cherry flavour, with some mint, and spice. The finish is fresh and lifted.

Wineries/Vineyards

The second-oldest Shiraz vineyard in Eden Valley, the **Poonawatta** vineyard, is located just north of Craneford. The vines of this 0.8 hectare benchmark vineyard were planted in 1880. The soil is sandy loam and skeletal rock. The Holt family, who bought the vineyard in 1966, have restored 24 rows, each of 38 vines.

Poonawatta has produced The 1880 Shiraz as their flagship wine since 2002, and it has characteristics typical of this area. The age of the vines is responsible for quite an intense wine, with its black fruit flavours due to the small berry size of the grapes. There are spice and savoury notes and soft tannins on the finish. Poonawatta is committed to expressing the characteristics of the vineyard, and the focus is the land and the vines. The 1880 Shiraz and all other Poonawatta wines are made at Kaesler's using typical small-scale production processes. Poonawatta uses French oak, up to two-thirds new, for The 1880 Shiraz.

Poonawatta also bottles another Shiraz from their vineyard: The Cuttings, from cuttings taken from the original vines in 1982. Another interesting Shiraz is the Four Corners Shiraz, which is a blend from the four different corners of Eden Valley.

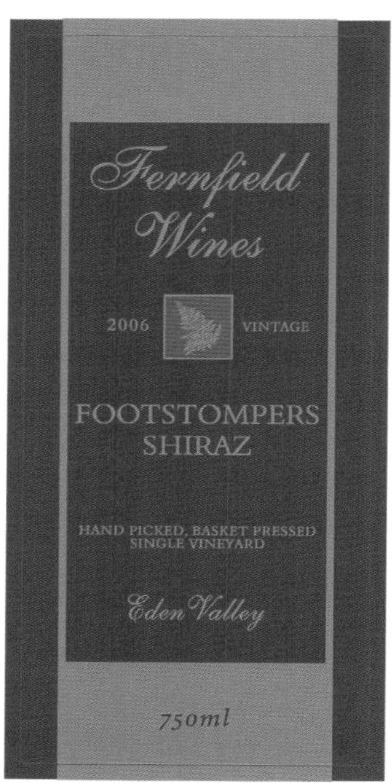

Fernfield Wines' cellar door near Eden Valley.

The cellar door of **Fernfield Wines** lies 1 kilometre east of the town of Eden Valley. The tasting room is in the Rushlea Homestead, which was built by William Lillecrapp, the founder of the town of Eden Valley, in 1866. The winery, set up in 2002, is now in the hands of the sixth generation of the family. Most grapes are sold to larger wineries, but Fernfield bottles two Shiraz labels of their own: the Pridmore and the Footstompers Shiraz.

This is a small operation, where absolutely everything, including the vineyard management, is done by hand and only by four family members (except for the bottling). The winemaking process is traditional, with hand picking, hand plunging, long fermentation and basket pressing. The main difference between the two Shirazes is the oak treatment. The Footstompers is aged in new French oak, the Pridmore 50 per cent in new French, 50 per cent in new American oak.

Between Eden Valley and Springton lies the Avon Brae Vineyard, which grows the fruit for **Eden Hall Wines**. Cherry and red fruit flavours, combined with spice and pepper, are typical for Shiraz from this area. At the southern end is Springton, with grey–brown podzolic soil. The main cellar door here is Poverty Hill.

High Eden

The sub-region of High Eden is in the west of Eden Valley. It was pushed by David Wynn, the founder of Mountadam, as a special place due to its high altitude: 400–600 m. The soil is special, too: yellow sandstone, mixed with quartz rocks. The vineyards here are mainly planted to Riesling and other white wine varieties.

Wineries/Vineyards

Mountadam and **Smallfry** make Shiraz from High Eden fruit. It tends to have less palate weight than other wines from the Barossa, but instead it has vitality, lengthy elegant fruit flavours, and some spice.

Mountadam is unusual in a number of ways. The winery is situated on top of a hill, whereas most other Barossa wineries are built in the valleys or on plateaus. It has a large landholding – 1000 hectares – of which 75 hectares are now planted. The wines are a pure expression of the estate and come in three ranges: the Barossa Range, the Eden Valley Range and, at the top of the pyramid, the High Eden Range. Fruit for the Patriarch Shiraz is grown at an elevation of 520 m, on a rocky quartz outcrop in one of the coolest winegrowing locations in Australia.

Mountadam's objective for the Patriarch Shiraz is that it should be a Shiraz in tune with the *terroir* of the sub-region. The fruit is hand pruned and hand picked and the wine is fermented in 100 per cent new oak. It is not high in alcohol, and balances fruit, spice and tannin.

Tasting profile

Generally speaking, the Shiraz from Eden Valley that has the biggest palate weight comes from the north, the Moculta area. As one travels further south, the palate weight reduces, acid levels increase, with the lightest and most lifted wines found in the High Eden area. Similarly, the palate of the northern wines includes mocha flavours. The further south one travels, the fresher the wines become, and red fruit flavours take over from dark fruit.

Eden Valley is often called a cool climate area, but the Shiraz generally does not display the intense spicy and peppery notes of cool climate Shirazes from Victoria. Fruit characters and aromatics dominate, and the wines are often elegant, and with fine tannins. Many of the vines are picked quite late. They are surprisingly alcoholic.

My 21 Favourite Barossa Shirazes

A list of favourites can be established in a number of ways: the ones that I would score the highest points, those that have the best consistency over many years, those that age best, those that deliver best value for money.

My list is dominated by fine wines which express the Barossa *terroir* particularly well. I have grouped them into a number of categories.

Icon wines

These are wines which deliver outstanding characteristics over many years and are unique in the world.

- Henschke: Hill of Grace
- Chris Ringland: Chris Ringland Shiraz

Full-bodied Barossa Shiraz

These are high quality wines which deliver the 'typical' Barossa Shiraz flavour and structure profile. In some, power dominates; others are more elegant. They share a strong tannin profile.

- Barossa Valley Estate: E&E Black Pepper Shiraz
- Burge Family Winemakers: Draycott Shiraz
- Elderton, Command Shiraz
- Henschke: Mount Edelstone Shiraz
- Kalleske: Greenock Shiraz
- Rolf Binder: Hanisch Shiraz
- Rockford: Basket Press Shiraz

Very ripe Shiraz

These wines are similar to the last group, but they push the envelope of ripeness a bit further. They are flagship wines of their respective wineries. Drinking a second or third glass can be a challenge.

- Kaesler: Old Bastard Shiraz
- Kalleske: Johann Georg Shiraz
- Torbreck: RunRig Shiraz Viognier

Aromatic Shiraz

These wines focus on aromatics and elegance, while still carrying the weight expected of a Barossa Shiraz.

- Grant Burge: Meshach Shiraz
- St Hallett: Old Block Shiraz
- Yalumba: The Octavius

Rhône blends

Some outstanding Shiraz is made where Shiraz is part of a blend, in particular with Grenache and Mataro/Mourvèdre.

- Charles Melton: Nine Popes
- Spinifex: Esprit
- Torbreck: The Steading

Value for money

The quality of these wines is very satisfying, and they come at a reasonable price.

- Fernfield: Pridmore Shiraz
- Teusner: Joshua Grenache Mataro Shiraz
- Torbreck: Woodcutter's Shiraz

Conclusion

10

The previous chapters describe in some detail the *terroir* of the Barossa sub-regions and the typical flavour and structure profiles of Shiraz wine. If I describe wine simply on two dimensions, the palate weight and the tannin structure, one can plot the typical profile of each sub-region on a matrix. This is, of course, simplistic, as it ignores the specific fruit and secondary flavours, but it allows some generalisations.

The dimension of palate weight includes fruit concentration, from light to strong; ripeness, from low to high; body, from light to full; and mouthfeel. The dimension of tannin structure goes from light to silky and fine-grained to strong and coarse-grained.

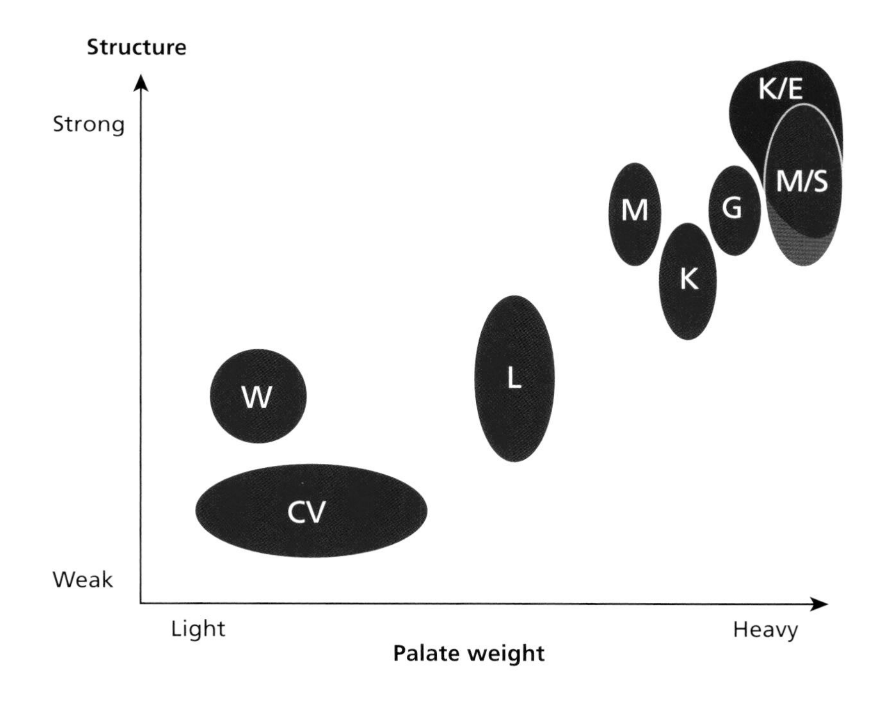

Flavour and Structure Profile of Barossa Sub-regions

CV = Central Valley W = Williamstown L = Lyndoch

M = Moppa K = Kalimna G = Greenock

M/S = Marananga/Seppeltsfield K/E = Koonunga/Ebenezer

Gomersal has a lot of variation

Eastern Slopes: not enough samples to classify

One can debate the exact placings, of course, but the general conclusion is fairly clear: many sub-regions of the Barossa Valley are located in the heavy palate weight, strong structure area of the matrix – the big Barossa. On the other hand, there are significant areas where it is typical to produce lighter weight wines with less tannin structure.

I make no comments about causal relationships between *terroir* and wine. Given that there is a debate about the influence *terroir* has, it would be presumptuous to develop a complete argument about this relationship. However, there are some standout features in a number of the sub-regions and in their grapes.

The Koonunga/Ebenezer sub-region, for example, is characterised by hot temperatures, but has high water-holding capacity due to its ironstone layer. Is this combination responsible for the big and often coarse-grained tannins?

Chocolate and mocha flavours tend to occur in areas where ironstone is present. Is there more than water and sun being picked up by the vines?

The extraordinary ripeness of vines in Marananga/Seppeltsfield may be due to the amount of sunshine, particularly on vineyards with eastern as well as western slopes. Are the fine and silky tannins from this area due to the very old schist rock in the area? A similar relationship may exist in parts of Eden Valley.

Deep sand is a feature in parts of the Barossa, in particular Kalimna, Vine Vale and parts of Lyndoch. The wines from these regions distinguish themselves by the purity and depth of their fruit characteristics and their lifted aromas.

At a minimum, I hope this book demonstrates the importance of *terroir*, and its impact. While we are far from being able to assess this as an exact science, we should also keep in mind that the history of fine table wine in the Barossa is not more than 25 years old, with a few exceptions.

I also feel that there is enormous potential for improvement in many vineyards. If the resources that go into the Hill of Grace vineyard or Penfolds' Kalimna vineyard, for example, could be applied to many other vineyards, the wine quality and uniqueness of the Barossa would increase dramatically across the board.

I believe that sub-regionality is meaningful in the Barossa, and I hope this book makes a contribution to understanding and appreciating the uniqueness of this extraordinary place and its wines.

Next steps

11

This is the first book to explore the sub-regionality of the Barossa in a systematic fashion. However, there are a number of obvious shortcomings:

1 The book is based on written material, interviews and tastings by the author. While most winemakers agreed to be interviewed, some did not. Also, while I have gained tasting impressions over a 25-year period, my systematic tastings took place only in 2009 and 2010. This is a snapshot of time and the specific weather patterns at that time may have led to non-typical results. This publication may open the debate more, and more people may come forward with their experiences. Wine Barossa has a systematic research program underway which will deliver controlled tasting results over at least five years.

2 It is difficult to be scientific about the boundaries of the sub-regions. A wider debate may lead to amended propositions.

3 The development of a group of what I have called benchmark vineyards is rudimentary. I consider this an important aspect of explaining sub-regionality and of the impact of *terroir.* Over time 'vineyard secrecy' will reduce and this work can be carried out more fully. I hope that eventually it will produce something similar to the *premier cru* and *grand cru* classifications in Europe.

4 This work is based on Shiraz only. Other important grapes are grown in the Barossa and should be incorporated. The freedom to grow what one pleases makes this task more difficult, but also more exciting than it would be in the controlled areas of Europe.

5 The information on Eden Valley needs to be more detailed. Once other grape varieties are included, this will become an easier task.

I hope that continued debate and research will increase clarity and improve the validity of the sub-regionality concept in the Barossa. If this is the case, I hope to include such improvements in a future edition of this book.

Bibliography

Allen, Max: *The Future Makers*, Hardie Grant Books, 2010
Australian Geographic Indication: Barossa Valley Region, 1997
Australian Geographic Indication: Eden Valley Region, 1997
Caillard, Andrew: *The Rewards of Patience*, 6th Edition, Allen & Unwin, 2008
Caillard, Andrew: *Beauty and Balance – The Razor's Edge,* Presentation at Nelson, NZ, February 2010
Carter, Felicity: Peter Lehmann: Vintage Barossa, Meininger's Wine Business International, 10.4.2008
Fairburn, W.A.: *The Geology of the Barossa Valley*, Primary Industries and Resources SA
Farmer, David: *History of the Barossa Valley and its Landscapes*, Wine Barossa Media and Trade Day, 2008
Farmer, David: How Does Soil and Rocks Influence the Taste of Wine? Glug.com.au, 19.5.2010
Fuller, Peter & Walsh, Brian: Barossa. Vintage Classification, Barossa Wine and Tourism Association, 1999
Girgensohn, Thomas: Subtleties of the Barossa Terroirs, *Winestate*, March/April 2011
Gladstones, John: *Wine, Terroir and Climate Change*, Wakefield Press, 2011
Goode, Jamie: Terroir Revisited: Towards a Working Definition, Wineanorak.com
Goode, Jamie: Mechanisms of Terroir, Wineanorak.com
Gregory, G.R.: Development and Status of Australian Viticulture, in B.G. Coombe and P.R. Dry, *Viticulture, Volume 1.*, Adelaide, Australian Industrial Publishers, 1988
Halliday, James: *Australian Wine Companion*, 2011 Edition, Hardie Grant Books, 2010
Jefford, Andrew: *The New France*, Mitchell Beazley, 2006
Jefford, Andrew: Regionality and its Myths, Andrewjefford blog, 23.10.2009
Jefford, Andrew: Falling in Love Again, Andrewjefford blog, 11.10.2009
Langton's: The Australian Cult Wines Market, langtons.com.au
Lofts, Graeme: *Heart & Soul, Australia's First Families of Wine*, John Wiley & Sons, 2010
Macarthur, William: *Letters on the Culture of the Vine, 1844*, reproduced on *Hortus Camdenensis* by Colin Mills, 2010
Nicholas, Phil: *Soil, Irrigation and Nutrition*, Wintau, 2004
Norrie, Philip: Who Imported the First Documented Shiraz Vines into Australia, The Wine Doctor
Northcote, K.H. et al.: *Soils and Land Use in the Barossa District*, 4 volumes, CSIRO, 1954–1959
Pitte, Jean-Robert: *Bordeaux/Burgundy*, University of California Press, 2008
Stelzer, Tyson & Dodd, Grant: *Barossa wine traveller*, Wine Press, 2009
Ward, Ebenezer: *The Vineyards and Orchards of South Australia*, Adelaide, 1862
White, Philip: Barossa Finally Gets Rocks in its Head, Drinkster Blog, 30.1.2010
Wine Barossa: Barossa Old Vine Charter, Notes, 16.4.2009
Wine Barossa: Barossa Sub-regional Shiraz Tasting, Notes, 27.7.2009

Winery Materials:
- Henschke: 1868–2008, *The Leader*, 10.9.2008
- Rockford: *The Rockford Rag Newsletter* 1985–2010
- Torbreck: Brochures on vineyards and growers
- Yalumba: *Distinguished Vineyards*, 2010/2011

Index

This index does not list names of wines, which are subject to frequent change. A few icon wine names are included. You may also need to check under different entries to find people, winery names, vineyard names, etc.

Wakefield Press is an independent publishing and
distribution company based in Adelaide, South Australia.
We love good stories and publish beautiful books.
To see our full range of books, please visit our website at
www.wakefieldpress.com.au
where all titles are available for purchase.